Beyond Resource Wars

Global Environmental Accord: Strategies for Sustainability and Institutional Innovation
Nazli Choucri, series editor

A complete list of books published in the Global Environmental Accord series appears at the back of this book.

Beyond Resource Wars

Scarcity, Environmental Degradation, and International Cooperation

Edited by Shlomi Dinar

The MIT Press
Cambridge, Massachusetts
London, England

For information about special quantity discounts, please email <special_sales@ mitpress.mit.edu>.

This book was set in Sabon by Toppan Best-set Premedia Limited. Printed and bound in the United States of America.

Library of Congress Cataloging-in-Publication Data

Beyond resource wars : scarcity, environmental degradation, and international cooperation / edited by Shlomi Dinar.
p. cm. — (Global environmental accord : strategies for sustainability and institutional innovation)
Includes bibliographical references and index.
ISBN 978-0-262-01497-7 (hardcover : alk. paper) — ISBN 978-0-262-51558-0 (pbk. : alk. paper)
1. Environmental policy—International cooperation. 2. Environmental management—International cooperation. 3. Environmental protection—International cooperation. I. Dinar, Shlomi.
JZ1324.B49 2011
333.7—dc22

2010017735

10 9 8 7 6 5 4 3 2 1

Contents

Series Foreword

A new recognition of profound interconnections between social and natural systems is challenging conventional constructs and the policy predispositions informed by them. Our current intellectual challenge is to develop the analytical and theoretical underpinnings of an understanding of the relationship between the social and the natural systems. Our policy challenge is to identify and implement effective decision-making approaches to managing the global environment.

The series Global Environmental Accord: Strategies for Sustainability and Institutional Innovation adopts an integrated perspective on national, international, cross-border, and cross-jurisdictional problems, priorities, and purposes. It examines the sources and the consequences of social transactions as these relate to environmental conditions and concerns. Our goal is to make a contribution to both intellectual and policy endeavors.

Nazli Choucri

Preface

This book was the product of a workshop titled Reflections on Resource Scarcity and Degradation: Conflict, Cooperation, and the Environment, which took place at Florida International University (FIU) as part of the Ruth K. and Shepard Broad Educational Series. Inspired by the rich literature (both policy and academic) that considers the relationship between resource scarcity and international conflict, the workshop's goal was to explore how scarcity, environmental degradation, and environmental change may also be a catalyst for cooperation across a number of natural resources and transboundary environmental issues.

The workshop, which was convened on November 17, 2006, and sponsored by the Shepard Broad Foundation Inc. and the Jack D. Gordon Institute for Public Policy and Citizenship Studies, brought together a team of scholars, and originally featured the topics of climate change as well as freshwater, fisheries, and oceans pollution. Additional sessions considered the general relationship between scarcity and cooperation, the linkage between international negotiation and environmental cooperation, and the role of environmental cooperation in promoting peace. The papers presented at the workshop constituted the initial chapters for this edited volume. Other chapters were subsequently solicited from experts in the field based on the theme of the workshop.

We are grateful to a number of individuals who provided invaluable feedback on the individual chapters or the workshop in general. They include Scott Barrett, Dag Herald Claes, Geoff Dabelko, Mark Giordano, Peter Jacques, Gordon Munro, David Simpson, Gunnar Sjöstedt, Marvin Soroos, John Tilton, and Kenneth Wilkening. We are also grateful to the three anonymous reviewers for their comments and suggestions on the manuscript. Likewise, we would like to thank Clay Morgan and the entire MIT editorial team for all their help.

Our gratitude also goes to Jan Solomon and Monalisa Gangopadhyay (two hardworking graduate students) for providing logistical support for the workshop. Several FIU faculty and staff also deserve special thanks for their support of the workshop, including David Bray, John Clark, François Debrix, Sarah Mahler, Michael McClain, Brian Fonseca, George Philippidis, Lisa Prügl, John Stack, and Mark Szuchman.

Contributors

J. Samuel Barkin is Associate Professor of Political Science at the University of Florida. He is the author of *International Organization: Theories and Institutions* (2006), and a coeditor of *Anarchy and the Environment: The International Relations of Common Pool Resources* (1999).

Elizabeth R. DeSombre is Frost Professor of Environmental Studies and Professor of Political Science. Her recent books include *Domestic Sources of International Environmental Policy: Industry, Environmentalists, and U.S. Power* (2000) and *Flagging Standards: Globalization and Environmental, Safety, and Labor Regulations at Sea* (2006).

Shlomi Dinar is Associate Professor in the Department of Politics and International Relations at Florida International University. He is a coauthor of *Bridges over Water: Understanding Transboundary Water Conflict, Negotiation, and Cooperation* (2007) and the author of *International Water Treaties: Negotiation and Cooperation along Transboundary Rivers* (2008).

Christopher J. Fettweis is Assistant Professor in the Department of Political Science at Tulane University. His recent publications include *Losing Hurts Twice as Bad: The United States in the Aftermath of Iraq* (2008) and *Dangerous Times: The International Politics of Great Power Peace* (2010).

Gabriela Kütting is Associate Professor in the Department of Political Science at Rutgers University at Newark. She is the author of *Globalization and the Environment: Greening Global Political Economy* (2004) and *The Global Political Economy of Environment and Tourism* (2010), and a coeditor of *Environmental Governance: Power and Knowledge in a Local-Global World* (2009).

Robert Mendelsohn is Edwin Weyerhaeuser Davis Professor at Yale University's School of Forestry and Environmental Studies. His area of interest is resource economics, with a special emphasis on valuing the environment. Some of his recent books with other coauthors and coeditors include *The Impact of Climate Change on Regional Systems* (2006), *Climate Change and Agriculture in Africa: Impact Assessment and Adaptation Strategies* (2008), and *Climate Change and Agriculture: An Economic Analysis of Global Impacts, Adaptation, and Distributional Effects* (2009).

G. Kristin Rosendal is Senior Research Fellow at the Fridtjof Nansen Institute in Lysaker, Norway. She is the author of *The Convention on Biological Diversity and Developing Countries* (2000).

Miranda A. Schreurs is Director of the Environmental Policy Research Centre and Professor of Comparative Politics at the Freie Universität in Berlin. She is the author of *Environmental Politics in Japan, Germany, and the United States* (2002), and a coeditor of *The Environmental Dimension of Asian Security: Conflict and Cooperation in Energy, Resources, and Pollution* (2007) and *Transatlantic Environment and Energy Politics: Comparative and International Perspectives* (2009).

Deborah J. Shields is Affiliate Faculty in the Department of Economics at Colorado State University and Visiting Professor in the Department of Land, Environment, and Geoengineering at Politecnico di Torino. Prior to her retirement, she was Principal Mineral Economist for the U.S. Department of Agriculture in the Forest Service, Research, and Development Division, where she directed the agency's energy and mineral economics along with its mineral policy research programs. She is a coauthor of *Sustainable Mineral Resource Management and Indicators: Case Study Slovenia* (2004).

Slavko V. Šolar is a Mineral Resource Geologist with the Geological Survey of Slovenia, in Ljubljana, Slovenia. He is a coauthor of *Sustainable Mineral Resource Management and Indicators: Case Study Slovenia* (2004).

I

Introduction

1

Resource Scarcity and Environmental Degradation: Analyzing International Conflict and Cooperation

Shlomi Dinar

The relationship among resource scarcity, degradation, and conflict has received a great deal of attention in both the international relations and environment and security literature. Both theoretical and empirical works have considered, either directly or indirectly, the environmental and natural resource roots of interstate conflict (Myers 1993; Tir and Diehl 1998; Homer-Dixon 1999; Matthew 1999). In its extreme yet rare form, the relationship between conflict and the environment has also been epitomized in the "resource wars" argument (Barnet 1980; Mandel 1988; Bullock and Darwish 1993; Baechler 1998; Klare 2001).

Past studies, critical of the relationship between resource scarcity, environmental degradation, and conflict, have largely argued that exogenous factors such as ingenuity, second-order resources, and trade often play a role in mitigating interstate violence over scarce resources (Simon 1989; Allan 2001). Yet the part that scarcity and degradation may in fact perform in fostering interstate cooperation, and consequently reducing both violent and political conflict, has received little attention (Deudney 1991, 1999).

This volume asserts that while resource scarcity and environmental degradation may well constitute sources of conflict, political dispute, and mismanagement between states, they may also be the impetus for cooperation, coordination, and negotiation between them. While the volume recognizes both sides of the resource scarcity and environmental degradation coin, the cooperative relationship is of particular interest and scrutiny. Indeed, conflict frequently motivates cooperation, and resource scarcity and environmental degradation are important elements of this relationship.

Generally, the authors in this volume maintain that increasing scarcity and degradation induce cooperation across states. To that extent, we provide a different perspective than that of the resource wars argument made with regard to particular natural resources such as oil, freshwater, minerals, and fisheries. Yet beyond this claim, the volume systematically explores the intricacies and nuances of this scarcity and degradation contention across a set of additional resources and environmental problems, which may merely motivate political conflicts such as climate change, ozone depletion, oceans pollution, transboundary air pollution, and biodiversity conservation. In particular, and in line with the collective action school, the volume investigates the notion that as scarcity and degradation worsen, interstate cooperation becomes difficult to achieve since it may be too costly to manage the degradation or there is simply too little of the resource to share (Ostrom 2001). Similarly, low levels of scarcity may depress cooperation as there is less urgency to organize and coordinate. Scarcity and degradation levels, in other words, should matter in explaining the intensity of cooperation.

While it is logical to associate the term "resource scarcity" with certain issue topics (e.g., oil and minerals) and the term "environmental degradation" with a slightly different set of issue topics (e.g., oceans pollution and transboundary air pollution), these two thematic labels are quite complementary. Environmental degradation often reduces the quantity or quality of the resource in question, thereby contributing to its scarcity. For example, "air pollution in a city degrades the quality of the air and changes an unlimited public good (clean air) into a scarce one." Likewise, the "pollution of a river . . . reduces the quality of the water; but it can also be interpreted as reducing the quantity of clean water, and therefore contributing to increased scarcity" (Gleditsch 1998, 387). To the degree that the two terms are not necessarily mutually exclusive, they are (for the most part) used interchangeably throughout this volume, when appropriate.

Recognizing that scarcity and degradation alone cannot explain the evolution of cooperation, the volume also considers other crucial factors for understanding the intricacies of interstate coordination. Although each topic boasts its own set of ancillary factors for explaining cooperation, this book investigates how asymmetries across countries (geographic, economic, and political) affect negotiation and cooperation (Botteon and Carraro 1997). To that extent, the volume looks at how respective asymmetries are managed to encourage interstate cooperation or environmental treaty making, in particular across transboundary environmental issues. While the asymmetries ascribed to the various issues help to explain the challenges to cooperation, they are directly interrelated with the book's theme on scarcity and degradation. In other words, countries that value the same resource differently or have varying abilities to deal with an environmental problem implicitly perceive scarcity and degradation in a divergent fashion. This does not necessarily mean that cooperation can't be achieved. Rather, such differences may need to be offset in order to encourage interstate coordination. The nature of cooperation may also be affected by different levels of scarcity as well as the asymmetries that usually influence how degradation and scarcity are perceived.

The volume's theme and hypotheses draw on scholars from disciplines and fields such as international relations, economics, and political science. Utilizing these disciplinary perspectives enables a more comprehensive and robust investigation of our scarcity-cooperation contention. A multidisciplinary method likewise provides a forum for better understanding important political, economic, legal, social, and cultural aspects of each natural resource or transboundary environmental issue along with the relationship to the evolution of international cooperation.

Scarcity, Degradation, and Conflict

Inspired by neo-Malthusian thinking, several studies have argued that resource scarcity and environmental degradation are key factors for understanding national and international insecurities in the form of violent conflict (Renner 1999). In their most acute form, such studies have considered the increased likelihood of resource wars among states. James Orme (1997, 165–167), for example, has generally asserted that the effects of climate change, global water scarcity, food security, and increased industrialization and population growth in the developing world could likely bring a fundamental alteration in the basic conditions of international politics. As a result, like in a Hobbesian world, states may well resort to force and war as they compete for scarce resources. This claim finds general support among realist thinkers who claim that tensions and conflict are more likely as countries attempt to reduce their dependence on other countries (Waltz 1979, 106, 154–155).

Lateral pressure theory, developed by Nazli Choucri and Robert North (1975, 1989), generally supports this claim theorizing that rivalry and conflict in the international system is partly guided by domestic growth and expansion and subsequent competition for resources and markets. Although the theory does not preclude peaceful resolutions to interstate competition for resources, it attests that countries may meet growing demands for resources by acquiring those necessary goods outside their own borders, which may be a source of conflict and war. Richard Ullman (1983, 139–140) has argued that "conflict over resources is likely to grow more intense as demand for some essential commodities increases and supplies appear more precarious." While focusing mostly on the relationship between resource scarcity and intrastate violence, Thomas Homer-Dixon (1999, 139) has contended that if a war was to be sparked over any renewable resource, it would be freshwater—citing the conflict-eliciting upstream/downstream configuration and power dynamics of the Nile River. Shy of an actual all-out war in the modern era, scholars have enumerated the militarized skirmishes that have been sparked over shared water resources. An often-cited example is the exchange of fire between Syria and Israel over Syria's "all-Arab" plan to divert the Jordan River headwaters (Seliktar 2005).

With much less frequency, international fisheries have also been discussed in the context of resource wars (Peterson and Teal 1986). A favorite near-armed example is that of the so-called Cod War of 1972–1973 between Iceland and the United Kingdom. In this case, Iceland unilaterally extended its coastal fishing rights, protecting access to what it considered a resource vital to its economy. The United Kingdom, which was also fishing in these waters and refused to recognize this extension, saw their trawlers driven away by Icelandic gunboats. In turn, this resulted in the arrival of the British Navy. The so-called Turbot War between Spain and Canada is another, more recent instance of a dispute, and one that actually involved live fire (Soroos 1997a; Barkin and DeSombre 2000). In 1995, a Spanish fishing boat was located outside the two-hundred-mile limit. The Spanish trawler was spotted by a Canadian patrol boat, which fired warning shots. While the tensions between the two countries eventually subsided, Canada had accused Spain of undermining conservation efforts.

Nonrenewable resources, as opposed to the above-mentioned renewable resources, have been touted as more prone to violent behavior between states, since they can be quickly utilized to build and fuel the military (Homer-Dixon 1999, 138–139). In this context, Michael Klare has claimed that access to oil could be a casus belli, focusing on the Persian Gulf, Caspian Sea, and South China Sea. In most of these cases, argues Klare (2001, 53, 57), the parties have yet to agree on a plan to divide up the spoils, making the situation potentially chaotic. Writing about the Middle East and Persian Gulf in the 1970s, Choucri with Vincent Ferraro (1976, 148) conjectured that the large-scale flow of armaments into the area could well increase the probability of local conflict and instability, consequently eliciting a response from oil-importing countries. Along somewhat similar lines, Neha Khanna and Duane Chapman (2010) have claimed that arms imports from the United States and its allies to the Persian Gulf have actually contributed to stable oil markets, at least between 1989 and 1999.

Arthur Westing's seminal edited volume *Global Resources and International Conflict* (1986a) seconds oil's strategic importance. One of the chapters on oil considers the timing of Germany's invasion of Russia in World War II as a consequence of Germany's lack of access to Middle Eastern oil supplies (Arbatov 1986, 24). Minerals are also discussed in the context of armed interstate conflict in Westing's work. Territorial rivalries over the iron-rich Lorrain region and conflicts over colonies or spheres of influence in resource-rich Africa played an important role in World War I (Westing 1986b, 204).

Beyond those resources (e.g., freshwater, oil, minerals, and fisheries) hypothesized to ignite violence between states or shown to have some role in fomenting war, other issues of environmental concern analyzed in this volume include ozone depletion, oceans pollution, biodiversity loss, and climate change. Surely considering these particular environmental topics as a casus belli would border on sensationalism.[1] Pertaining especially to climate change, however, scholars have considered how the effects of global warming may indirectly undermine the legitimacy of governments, affect human health and food security, and induce migration (Barnett 2003; Nordås and Gleditsch 2007; Reuveny 2007). By extension these studies have argued that competition and violence may ensue. In fact, empirical studies have found moderate support for a connection between the effects of climate change and civil conflict (Raleigh and Urdal 2007).

It is likely that the violent conflict hypothesized by studies, as a result of climate change and its effects, is unique to the intrastate level (i.e., militarized internal conflicts) rather than the interstate one (i.e., militarized state-to-state interactions). Yet bearing in mind the consequences of climate change and the other environmental issues enumerated above to regional and international security, in a broad sense, would be germane, since the issues relate to a state's and region's economic well-being and environmental sustainability (Gleick 1989; Soroos 1997b; Benedick 1998, Nautilus Institute and Center for Global Communications 2000; Jacques 2002, 2006; McNeely 2005; IPCC 2007). Consequently, regional and international security, broadly defined, may also be affected to the extent that a nonviolent political dispute regarding an environmental issue ensues and remains unresolved (Mathews 1989, 174; Esty 1999, 304, 308).

Predicting the occurrence of formal militarized disputes among states over scarce natural resources (particularly water, oil, minerals, and fisheries) is a somewhat futile exercise as it is difficult to ascertain what the future holds. By one account, though, the period 1901–1986 has seen only nine interstate wars with a specific connection to such resources (Westing 1986b, 204–210).[2] In contrast, the periods 1920–1944 and 1946–2010 have seen the signing of around 108 and 2,719 agreements, respectively, pertaining to those same resources (United Nations Treaty Collection n.d.).[3] The United Nations Treaty Collection also reports a total of about 1,128 treaties pertaining to the environment, in particular, for the periods 1920–1944 and 1946–2010.[4] While surely these simple numerical comparisons do not debunk the possibility that a militarized violent dispute may occur over scarce resources, it is clear that exercises of cooperation are, in comparison, impressive in number.

This is not to suggest that conflicts of interest and political disputes do not take place over scarce resources or a degraded environment (Gleditsch 1998, 387). Indeed, relationships of complex interdependence may motivate tensions among states (Keohane and Nye 1977, 9–11). That being said, it is these disputes and political tensions that may likewise inspire coordination. It is thus important to understand why and how these issues motivate interstate cooperation and environmental treaty making. Resource scarcity and environmental degradation are at the root of explaining such phenomena.

Scarcity, Degradation, and Cooperation

Critics of the resource wars and environment-conflict conjectures have mostly argued from an ingenuity or second-order resources standpoint, among other perspectives.

The late economist Julian Simon (1981, 3–11, 345–348), for example, maintained that people are the "ultimate resource" and the effects of scarcity could be overcome through human ingenuity. Esther Boserup (1981, 3–28, 93–111), the agricultural economist, has seconded this idea, demonstrating that (prior to the Industrial Revolution) densely populated countries, where there would be more strain on the environment, were actually beacons of technological innovation. In other words, the necessity to cope with environmental scarcities—instigated by a large population—motivates invention.

A second-order resources perspective considers the availability of institutional capacities to deal with the social consequences of considerable physical scarcity or first-order resources (Ohlsson 1999; Ohlsson and Turton 2000).

Aside from a small number of works, the premise that resource scarcity and environmental degradation actually provide the impetus for cooperation and negotiation has not been developed in greater detail. Unquestionably, these works provide the rationale and starting point for this volume. That is, scarcity is the basic impetus for interstate cooperation. As the Scottish philosopher David Hume ([1739–1740] 1978, 494–498) postulated, the need for rules of justice is not universal. Such rules arise only under conditions of relative scarcity, where property must be regulated to preserve order in society.

Most notable of the few contemporary writings that directly link scarcity and degradation with international cooperation is Daniel Deudney's critical analysis of the scarcity-conflict contention. In the edited volume titled *Contested Grounds*, Deudney (1999, 203) argues that "analysts of environmental conflict do not systematically consider ways in which environmental scarcity or change can stimulate cooperation."[5] Others have likewise joined Deudney's assertion, challenging the environment-conflict thesis. Jon Barnett, for example, has even criticized more "reasoned" works that question the resource wars argument (as they pertain to raw materials) as not going far enough since they still see a real possibility for environmentally induced conflicts (say, over water). According to Barnett (2000, 274), such studies have nevertheless given the ontological priority to conflict over cooperation.[6] Albeit concentrating on the effects of resource scarcity on individual island states, Richard Matthew and Ted Gaulin (2001) also challenge the popular conflictual scenarios hypothesized in the literature, demonstrating how scarcity has actually led to cooperation.

Another important study considers the relationship between environmental cooperation and peacemaking (Conca and Dabelko 2002). The book tests the proposition that environmental cooperation may be a trigger for reducing tensions, broadening cooperation, and promoting peace on other political levels. While Ken Conca and Geoffrey Dabelko's volume considers a broader spectrum of cooperation, it goes beyond the traditional relationship linking environmental degradation and violent conflict, and tackles the issue of environmental coordination. To that extent, it is an inspiration for this particular volume.

While this volume seeks to more systematically consider the general relationship between resource scarcity, environmental degradation, and cooperation, it attempts to make another contribution to the literature.

In particular, the collective action school provides that additional rational for this book.

John Rawls has conjectured that when natural and other resources are abundant, schemes of cooperation become superfluous. Conversely, when conditions are particularly harsh, fruitful ventures break down. A situation of moderate (or relative) scarcity therefore provides a suitable impetus for action between parties (Rawls 1971, 127–128). Similarly, Elinor Ostrom and her colleagues have argued that for cooperation or successful governance to occur, "resource conditions must not have deteriorated to such an extent that the resource is useless, nor can the resource be so little used that few advantages result from organizing" (1999, 281; see also Dolšak and Ostrom 2003, 12–13). As Ted Gurr (1985, 53) concurs, "political constraints weigh heavily on what might be achieved collectively in the face of serious scarcity."

A related association claims that when the economic burden of dealing with an environmental externality increases, the incentives to creating interstate regulations are lower (Sprinz and Vaahtoranta 1994). In turn, abatement cost functions and the general ability to alleviate degradation—say, through technological innovation—are partly hampered by the severity of the degradation and its impacts (Barbier and Homer-Dixon 1996; Homer-Dixon 1999, 108; Chasek, Downie, and Brown 2006, 205). Consequently, degradation or scarcity should not have become so severe that it is too costly to manage, thereby making international coordination less likely.

In the context of the book's general investigation into the links between scarcity and cooperation, it is expected that cooperation may be inhibited at low and high levels of scarcity and degradation. In other words, there is less incentive to coordinate actions, since the resource is relatively plentiful or environmental harm is not believed to be serious enough. Similarly, cooperation is expected to be less likely when scarcity and degradation are at a high level, either because there is little of the resource to divide among the parties or because the environmental harm is relatively costly to manage.

Beyond Scarcity and Degradation: Additional Imperative Factors

Cooperation may be motivated by resource scarcity and environmental degradation, but it does not necessarily materialize because of these two important elements. This volume highlights a set of other factors crucial for understanding how cooperation is facilitated or mitigated. The general desire for stable markets and prices, in the context of lucrative resources such as oil and minerals, is hypothesized to lead to interstate cooperation (Mikdashi 1976, 118; Westing 1986c, 12; Arbatov 1986, 36; Hveem 1986, 81; Chapman and Khanna 2006, 511). The overall political environment and the level of influence of domestic forces in favor of cooperation are likewise argued to be significant factors (Putnam 1988; Milner 1997, 60–64; Moravcsik 1997). Finally, differences in the countries' political systems and structures may also play a role in facilitating or inhibiting cooperation (Gleditsch 1998; Leeds 1999; Martin 2000, 26, 47).

Beyond these factors, this volume is particularly interested by the manner in which environmental problems often arise among asymmetrical players (Faure and Rubin 1993, 23; Susskind 1994, 18–19). This book thus scrutinizes the difficulties such asymmetries may pose and investigates how they are overcome so as to facilitate cooperation. While the authors of this volume identify a number of asymmetries unique to their own chapters, several issues are discernible across the majority of the cases. They are discussed below.

The Relevance of Country Asymmetries to Understanding Conflict and Cooperation

The ramifications and effects of economic asymmetries on negotiation and cooperation are commonly discussed in the environmental politics and economics literature. As Michele Botteon and Carlo Carraro (1997, 27) argue, environmental problems are often characterized by large asymmetries across countries in terms of both the benefits received and the costs accrued from, say, abating pollution. In addition, poorer countries tend to have shorter shadows of the future (or high discount rates) with regard to the resource (Dasgupta and Mäler 1994, 4–5; Fairman 1996, 69; Compte and Jehiel 1997, 63). This may exacerbate an environmental problem and make cooperation more difficult to attain (Scott 1974, 842; Ostrom 1992, 299). Such countries may have higher propensities to pollute or simply prioritize more pressing issues over environmental protection (Keohane 1996, 3–4; Barkin and Shambaugh 1999a, 13; Barkin and Shambaugh 1999b, 178; Darst 2001, 39). To that extent, we expect richer states may have to provide incentives, such as compensation, if they wish to conclude an environmental agreement and minimize defection (Young 1994, 128, 132–133; Keohane 1996, 5; Raustiala and Victor 1998, 696; Underdal 2002b, 123; Barrett 2003, 335–351). Negative incentives or inducements, such as trade restrictions, can also be employed to offset asymmetries and encourage cooperation (Levy 1997; Barrett 2003, 310–315). Often, however, positive gestures make would-be international agreements more legitimate, contributing to their effectiveness (Connolly 1996, 334–335; Bodansky 1999, 603; Oakerson 1992, 52).

The discussion on economic asymmetries may logically relate to a more general assessment of power broadly defined. In other words, more powerful states in aggregate power terms may find themselves in a weaker position in relation to the environmental issues negotiated (Habeeb 1988; Zartman and Rubin 2000, 289). Issue-specific structural power therefore may effectively favor the otherwise-weaker state (in aggregate power terms), assuming the richer and more powerful country has a longer shadow of the future toward the environmental good in question. While, in line with hegemonic stability theory, mightier parties may be more inclined to take the lead in initiating environmental regimes, they are not necessarily able to exercise power over other states to their sole advantage. As Oran Young (1994, 135) attests, "Those countries in possession of structural power will often find that they can achieve more by using their power to make promises and offer rewards than they can by relying on threats and punishments."

Second, the different effects of the environmental externality or natural resource scarcity on a given country may also be of particular relevance in either prolonging or mitigating the dispute. Related to the discussion on economic asymmetries, a given party may have different propensities to, for example, accept pollution. In addition, the same environmental problem may have different effects across time on the same party. The less that countries are affected by a given environmental externality, the less urgency they will have in responding to that problem in a concerted fashion with other parties, which may be more affected (Young 1989, 354). By extension, the more countries are dependent on a particular resource, the more concerned they will be with its present and future viability. The dispute thus may either take on a prolonged state, or incentives from those states more affected by the externality or dependent more on the resource may have to be forthcoming to those parties less affected by the externality or less dependent on the resource.

Third, the transboundary nature of environmental externalities makes the effects of geographic asymmetries relevant. The directionality of particular environmental problems may not only exacerbate a dispute but also affect cooperation and negotiation (Weinthal 2002, 25; Giordano 2003, 371–372). Those countries that are situated upstream or upwind may have a strategic advantage over their neighboring states (Sprout and Sprout 1962, 366). In this instance, the externality is unidirectional (rather than reciprocal), which in general affects the downstream or downwind countries substantially more (Matthew 1999, 171). Naturally, when directionality issues are combined with the different economic characteristics of countries, several scenarios pertinent to conflict, cooperation, and bargaining power are possible in the case of both unidirectional and reciprocal externalities (Barkin and Shambaugh 1999b; Zartman and Rubin 2000, 289). Inducements and other bargaining strategies (such as compensation or issue linkage) may have to be considered in this scenario as well so as to offset inherent asymmetries and encourage cooperation.

While the asymmetries described above may have their exogenous effects on conflict, cooperation, and negotiation, they are also inextricably linked to our conception of scarcity and environmental degradation. To the extent that countries have different shadows of the future, are affected differently by the pollution or diminishing resource, and are differently situated along an environmental commons in terms of scale and space, such countries may have different conceptions of scarcity and degradation (Cooper 1989, 181). For example, since a downstream or downwind state is (all things being equal) typically more concerned with the effects of a unidirectional externality like pollution, such a state may perceive the degradation to be more serious, and the resource increasingly scarce, compared to its upstream or upwind neighbor. These nuances—that is, the lopsidedness in the interdependent relationship among parties in terms of scarcity and degradation—are also important for an exploration of the evolution of cooperation (Knorr 1975, 221–222; Keohane and Nye 1977, 9–11; Mandel 1988, 32). In other words, country asymmetries (affecting each party's perception of scarcity and degradation) make particular issues that much more "malignant" and increasingly difficult to solve (Underdal 2002a, 19).

The Chapters

Part II of the book investigates the global commons. In chapter 2, Robert Mendelsohn analyzes the case of climate change. He considers the stock of greenhouse gases in the atmosphere and the asymmetries among the respective countries as key to understanding why an optimal cooperative regime has not materialized. In chapter 3, Elizabeth DeSombre scrutinizes the case of ozone. She looks at the uncertainty that surrounded ozone depletion, the cooperation that nonetheless ensued, and then the deepening of that cooperation as the uncertainty regarding degradation and its causes abated. The underlying asymmetries among the parties are likewise important, especially as they elicited different bargaining tactics and eventually shaped the negotiated outcome. Chapter 4, by Kristin Rosendal, explores global biodiversity cooperation. The author examines how the physical loss of biodiversity, in addition to its economic value, gave rise to the Convention on Biological Diversity in the 1990s. The evolution of environmental norms at the time of the agreement is also discussed. Asymmetries between developing and developed countries are considered paramount for understanding how the benefits from cooperation were divided among the parties. Despite a seemingly impressive treaty, implementation of the agreement has become problematic. The uncertainty in the value of biodiversity and the effective abundance of the resource, at least in the short term, may help to explain the difficulties with implementation.

Part III of the book takes a regional approach to the analysis of scarcity and environmental degradation. In chapter 5, Miranda Schreurs considers how transboundary air pollution was framed as a problem, based on its effects, in North America, Europe, and East Asia, and the timing of the policy responses that followed. She also looks at some of the obstacles that slowed down the development of regimes, particularly in East Asia, including domestic influences and country asymmetries. Chapter 6, by Gabriela Kütting, explores the case of oceans pollution in the context of the Mediterranean Action Plan. Her analysis points to the increased capacity building and scientific cooperation among the regional players as a result of the sea's degradation. Yet she describes why relatively little action has taken place on the formal policy level, in light of the transborder effects of the pollution.

Part IV is dedicated to those stock resources that are generally seen, in both the academic and popular press, to be the main cause of future wars. In chapter 7, Samuel Barkin discusses three types of outcomes given the evidence of fish stock depletion: continued degradation, conflict, or cooperation. He argues that the particular outcome is often determined by the parties' differing shadows of the future toward the resource, which in turn are made up of several other country differences and asymmetries. In chapter 8, I consider the case of freshwater, with special emphasis on transboundary rivers. I look at the relationship between scarcity and cooperation, accounting for the large corpus of documented water agreements. While keeping domestic influences in mind, I also explore the various asymmetries that must be overcome to promote cooperation across specific property rights outcomes by analyzing specific international water agreements. Chapter 9, by Christopher Fettweis, investigates the case of oil. He studies three different regions with distinct contestation-inducing characteristics (by design, identical to those cited by Klare). In his analysis, he examines the relationship between consumer and producer states, including the behavior of great powers and regional players throughout history, assessing the costs and benefits of conflict and cooperation. Economically based incentives, in addition to other critical variables, are investigated in their relationship to cooperation. Chapter 10, by Deborah Shields and Slavko Šolar, examines the increased demand for minerals, and the overall relationship to short- and long-term scarcity, including real and perceived scarcity. Beyond sheer physical scarcity as a catalyst for conflict and cooperation, the authors consider other forms of scarcity. The authors tout additional cooperation-inducing factors such as market-, policy-, and consensus-based incentives. Chapter 11 concludes the volume by synthesizing the findings of the above chapters as they pertain to the conjectures presented. Further implications for theory and policy are also provided.

Notes

I would like to thank fellow volume contributors (and particularly Samuel Barkin) for their comments on this chapter before it was anonymously reviewed.

1. Peter Schwartz and Doug Randall (2003) discuss the possibility of climate change educing violent interstate conflict.

2. Westing actually identifies twelve incidents, but three of them are of an intrastate nature.

3. The search consisted of the subject terms "fishing and fisheries," "mineral resources," "mining," "petroleum," and "watercourses-water resources." Additional information was provided by Andrei Kolomoets, Legal Information Officer in the Treaty Section of the Office of Legal Affairs at the United Nations.

4. The search consisted of the subject term "environment." Other depositories of environmental agreements include Barrett (2003, 165–194) with about 290 multilateral agreements. The International Environmental Agreements Database Project out of the University of Oregon includes about 2,750 treaties. See <http://iea.uoregon.edu/page.php?file=home.htm&query=static> (accessed May 13, 2010).

5. For related writings, see Dokken 1997, 519–520, 533; Brock 1992, 99.

6. In this case, Barnett was specifically referring to Lipschutz and Holdren 1990.

References

Allan, Anthony. 2001. *The Middle East Water Question*. London: I. B. Tauris Publishers.

Arbatov, Alexander. 1986. "Oil as a factor in strategic policy and action: Past and present." In *Global Resources and International Conflict: Environmental Factors in Strategic Policy and Action*, ed. Arthur Westing, 21–54. Oxford: Oxford University Press.

Baechler, Günther. 1998. "Why environmental transformation causes violence: A synthesis." *Environmental Change and Security Project Report* (4): 24–44.

Barbier, Edward, and Thomas Homer-Dixon. 1996. "Resource scarcity, institutional adaptation, and technical innovation: Can poor countries attain endogenous growth." Project on Environment, Population, and Security Occasional Paper.

Barkin, Samuel, and Beth DeSombre. 2000. "Unilateralism and multilateralism in international fisheries management." *Global Governance* 6 (3): 339–360.

Barkin, Samuel, and George Shambaugh. 1999a. "Hypotheses on the international politics of common pool resources." In *Anarchy and the Environment: The International Relations of Common Pool Resources*, ed. Samuel Barkin and George Shambaugh, 1–25. Albany: State University of New York Press.

Barkin, Samuel, and George Shambaugh. 1999b. "Conclusions: Common pool resources and international environmental negotiation." In *Anarchy and the Environment: The International Relations of Common Pool Resources*, ed. Samuel Barkin and George Shambaugh, 176–198. Albany: State University of New York Press.

Barnet, Richard. 1980. *The Lean Years: Politics in the Age of Scarcity*. New York: Simon and Schuster.

Barnett, Jon. 2000. "Destabilizing the environment-conflict thesis." *Review of International Studies* 26 (2): 271–288.

Barnett, Jon. 2003. "Security and climate change." *Global Environmental Change* 13 (1): 7–17.

Barrett, Scott. 2003. *Environment and Statecraft: The Strategy of Environmental Treaty Making*. Oxford: Oxford University Press.

Benedick, Richard. 1998. *Ozone Diplomacy: New Directions in Safeguarding the Planet*. Cambridge, MA: Harvard University Press.

Bodansky, Daniel. 1999. "The legitimacy of international governance: A coming challenge for international environmental law." *American Journal of International Law* 93 (3): 596–624.

Boserup, Esther. 1981. *Population and Technological Change: A Study of Long-term Trends*. Chicago: University of Chicago Press.

Botteon, Michele, and Carlo Carraro. 1997. "Burden sharing and coalition stability in environmental negotiations with asymmetric countries." In *International Environmental Negotiations: Strategic Policy Issues*, ed. Carlo Carraro, 26–55. Cheltenham: Edward Elgar.

Brock, Lothar. 1992. "Security through defending the environment: An illusion?" In *New Agendas for Peace Research: Conflict and Security Reexamined*, ed. Elise Boulding, 79–102. Boulder, CO: Lynne Rienner.

Bullock, John, and Adel Darwish. 1993. *Water Wars: Coming Conflicts in the Middle East*. London: Victor Gollancz.

Chapman, Duane, and Neha Khanna. 2006. "The Persian Gulf, global oil resources, and international security." *Contemporary Economic Policy* 2 (4): 507–519.

Chasek, Pamela, David Downie, and Janet Brown. 2006. *Global Environmental Politics*. Boulder, CO: Westview Press.

Choucri, Nazli, with Vincent Ferraro. 1976. *International Politics of Energy Interdependence*. Lexington, MA: D. C. Heath.

Choucri, Nazli, and Robert North. 1975. *Nations in Conflict: National Growth and International Violence*. San Francisco: W. H. Freeman and Company.

Choucri, Nazli, and Robert North. 1989. "Lateral pressure in international relations: Concept and theory." In *Handbook of War Studies*, ed. Manus Midlarsky, 289–354. Boston: Unwin Hyman.

Compte, Olivier, and Philippe Jehiel. 1997. "International negotiation and dispute resolution mechanisms: The case of environmental negotiations." In *International Environmental Negotiations: Strategic Policy Issues*, ed. Carlo Carraro, 56–70. Cheltenham, UK: Edward Elgar.

Conca, Ken, and Geoffrey Dabelko, eds. 2002. *Environmental Peacemaking*. Washington, DC and Baltimore, MD: Woodrow Wilson Center Press and the Johns Hopkins University Press.

Connolly, Barbara. 1996. "Increments for the earth: The politics of environmental aid." In *Institutions for Environmental Aid: Pitfalls and Promise*, ed. Robert Keohane and Marc Levy, 327–365. Cambridge, MA: MIT Press.

Cooper, Robert. 1989. "International cooperation in public health as a prologue to macroeconomic cooperation." In *Can Nations Agree?* ed. Robert Cooper, Barry Eichengreen, C. Randall Henning, Gerald Holtham, and Robert Putnam, 178–254. Washington, DC: Brookings Institution Press.

Darst, Robert. 2001. *Smokestack Diplomacy: Cooperation and Conflict in East-West Environmental Politics*. Cambridge, MA: MIT Press.

Dasgupta, Partha, and Karl-Göran Mäler. 1994. "Poverty, institutions, and the environmental-resource base." *World Bank Environment Paper* 9: 1–74.

Deudney, Daniel. 1991. "Environment and security: Muddled thinking." *Bulletin of the Atomic Scientists* 47 (3): 22–28.

Deudney, Daniel. 1999. "Environmental security: A critique." In *Contested Grounds: Security and Conflict in the New Environmental Politics*, ed. Daniel Deudney and Richard Matthew, 187–219. Albany: State University of New York Press.

Dokken, Karen. 1997. "Environmental conflict and international integration." In *Conflict and the Environment*, ed. Nils Petter Gleditsch, 519–534. Dordrecht: Kluwer Academic Publishers.

Dolšak, Nives, and Elinor Ostrom. 2003. "The challenges of the commons." In *The Commons in the New Millennium: Challenges and Adaptation*, ed. Nives Dolšak and Elinor Ostrom, 3–34. Cambridge, MA: MIT Press.

Esty, Daniel. 1999. "Pivotal states and the environment." In *The Pivotal States: A New Framework for U.S. Policy in the Developing World*, ed. Robert Chase, Emily Hill, and Paul Kennedy, 290–314. New York: W. W. Norton and Company.

Fairman, David. 1996. "The global environmental facility: Haunted by the shadow of the future." In *Institutions for Environmental Aid: Pitfalls and Promise*, ed. Robert Keohane and Marc Levy, 55–87. Cambridge, MA: MIT Press.

Faure, Guy Olivier, and Jeffrey Rubin. 1993. "Organizing concepts and questions." In *International Environmental Negotiation*, ed. Gunnar Sjöstedt, 17–26. Newbury Park, CA: Sage Publications.

Giordano, Mark. 2003. "The geography of the commons: The role of scale and space." *Annals of the Association of American Geographers* 93 (2): 366–375.

Gleditsch, Nils Petter. 1998. "Armed conflict and the environment: A critique of the literature." *Journal of Peace Research* 35 (3): 381–400.

Gleick, Peter. 1989. "The implications of global climatic changes for international security." *Climatic Change* 15 (1–2): 309–325.

Gurr, Ted. 1985. "On the political consequences of scarcity and economic decline." *International Studies Quarterly* 29 (1): 51–75.

Habeeb, William. 1988. *Power and Tactics in International Negotiation: How Weak Nations Bargain with Strong Nations*. Baltimore, MD: Johns Hopkins University Press.

Homer-Dixon, Thomas. 1999. *Environment, Scarcity, and Violence*. Princeton, NJ: Princeton University Press.

Hume, David. [1739–1740] 1978. *A Treatise of Human Nature*. Oxford: Clarendon Press.

Hveem, Helge. 1986. "Minerals as a factor in strategic policy and action." In *Global Resources and International Conflict: Environmental Factors in Strategic Policy and Action*, ed. Arthur Westing, 55–84. Oxford: Oxford University Press.

IPCC (Intergovernmental Panel on Climate Change). 2007. *Synthesis Report*. Geneva, Switzerland.

Jacques, Peter. 2002. "Ocean security, sustainable development, and peace." *International Journal of Humanities and Peace* 18 (1): 86–89.

Jacques, Peter. 2006. *Globalization and the World Ocean*. Lanham, MD: Rowman and Littlefield.

Keohane, Robert. 1996. "Analyzing the effectiveness of international environmental institutions." In *Institutions for Environmental Aid: Pitfalls and Promise*, ed. Robert Keohane and Marc Levy, 3–27. Cambridge, MA: MIT Press.

Keohane, Robert, and Josephy Nye. 1977. *Power and Interdependence: World Politics in Transition*. Boston: Little, Brown.

Khanna, Neha, and Duane Chapman. 2010. "Guns and oil: An analysis of conventional weapons trade in the post cold war era." *Economic Inquiry* 48 (2): 434–459.

Klare, Michael. 2001. *Resource Wars: The New Landscape of Global Conflict.* New York: Metropolitan Books.

Knorr, Klaus. 1975. *The Power of Nations: The Political Economy of International Relations.* New York: Basic Books.

Leeds, Brett Ashley. 1999. "Domestic political institutions, credible commitments, and international cooperation." *American Journal of Political Science* 43 (4): 979–1002.

Levy, Jack. 1997. "Prospect theory, rational choice, and international relations." *International Studies Quarterly* 41 (1): 87–112.

Lipschutz, Ronnie, and John Holdren. 1990. "Crossing borders: Resource flows, the global environment, and international stability." *Bulletin of Peace Proposals* 21 (2): 121–133.

Mandel, Robert. 1988. *Conflict over the World's Resources: Background, Trends, Case Studies, and Considerations for the Future.* New York: Greenwood.

Martin, Lisa. 2000. *Democratic Commitments: Legislators and International Cooperation.* Princeton, NJ: Princeton University Press.

Mathews, Jessica. 1989. "Redefining security." *Foreign Policy* 68 (2): 162–177.

Matthew, Richard. 1999. "Introduction: Mapping contested grounds." In *Contested Grounds: Security and Conflict in the New Environmental Politics*, ed. Daniel Deudney and Richard Matthew, 1–15. Albany: State University of New York Press.

Matthew, Richard, and Ted Gaulin. 2001. "Conflict or cooperation? The social and political impacts of resource scarcity on small island states." *Global Environmental Politics* 1 (2): 48–70.

McNeely, Jeffrey. 2005. "Biodiversity and security." In *Human and Environmental Security: An Agenda for Change*, ed. Felix Dodds and Tim Pippard, 139–151. Sterling, VA: Earthscan.

Mikdashi, Zuhayr. 1976. *The International Politics of Natural Resources.* Ithaca, NY: Cornell University Press.

Milner, Helen. 1997. *Interests, Institutions, and Information.* Princeton, NJ: Princeton University Press.

Moravcsik, Andrew. 1997. "Taking preferences seriously: A liberal theory of international politics." *International Organization* 51 (4): 513–553.

Myers, Norman. 1993. *Ultimate Security: The Environment as the Basis of Political Stability.* New York: W. W. Norton and Company.

Nautilus Institute and Center for Global Communications. 2000. "Energy, environment, and security in northeast Asia: Defining a US-Japan partnership for regional comprehensive security." *ESENA Project Final Report.* Berkeley, CA: Nautilus Institute for Security and Sustainable Development.

Nordås, Ragnhild, and Nils Petter Gleditsch. 2007. "Climate change and conflict." *Political Geography* 26 (6): 627–638.

Oakerson, Ronald. 1992. "Analyzing the commons: A framework." In *Making the Commons Work: Theory, Practice, and Policy*, ed. Daniel Bromley, David Feeny, Margaret McKean, Pauline Peters, Jere Gilles, Ronald Oakerson, C. Ford Runge, and James Thomson, 41–59. San Francisco: Institute for Contemporary Studies Press.

Ohlsson, Leif. 1999. "Environment, scarcity, and conflict: A study in Malthusian concerns." PhD diss., University of Göteborg, Sweden.

Ohlsson, Leif, and Anthony Turton. 2000. "The turning of a screw: Social resource scarcity as a bottle-neck in adaptation to water scarcity." *Stockholm Water Front* 1: 10–11.

Orme, John. 1997. "The utility of force in a world of scarcity." *International Security* 22 (3): 138–167.

Ostrom, Elinor. 1992. "The rudiments of a theory of the origins, survival, and performance of common property institutions." In *Making the Commons Work: Theory, Practice, and Policy*," ed. Daniel Bromley, David Feeny, Margaret McKean, Pauline Peters, Jere Gilles, Ronald Oakerson, C. Ford Runge, and James Thomson, 293–318. San Francisco: Institute for Contemporary Studies Press.

Ostrom, Elinor. 2001. "Reformulating the commons." In *Protecting the Commons: A Framework for Resource Management in the Americas*, ed. Joanna Burger, Elinor Ostrom, Richard Norgaard, David Policansky, and Bernard Goldstein, 17–41. Washington, DC: Island Press.

Ostrom, Elinor, Joanna Burger, Christopher Field, Richard Norgaard, and David Policansky. 1999. "Revisiting the commons: Local lessons, global challenges." *Science* 284 (5412): 278–282.

Peterson, Susan, and John Teal. 1986. "Ocean fisheries as a factor in strategic policy and action." In *Global Resources and International Conflict: Environmental Factors in Strategic Policy and Action*, ed. Arthur Westing, 114–142. Oxford: Oxford University Press.

Putnam, Robert. 1988. "Diplomacy and domestic politics: The logic of two-level games." *International Organization* 42 (3): 427–460.

Raleigh, Clionadh, and Henrik Urdal. 2007. "Climate change, environmental degradation, and armed conflict." *Political Geography* 26 (6): 674–694.

Raustiala, Kal, and David Victor. 1998. "Conclusions." In *The Implementation and Effectiveness of International Environmental Commitments: Theory and Practice*, ed. David Victor, Kal Raustiala, and Eugene Skolnikoff, 659–707. Cambridge, MA: MIT Press.

Rawls, John. 1971. *A Theory of Justice*. Cambridge, MA: Belknap Press of Harvard University Press.

Renner, Michael. 1999. "Ending violent conflict." *Worldwatch Paper* 146: 1–71.

Reuveny, Rafael. 2007. "Climate change-induced migration and violent conflict." *Political Geography* 26 (6): 656–673.

Schwartz, Peter, and Doug Randall. 2003. *An Abrupt Climate Change Scenario and Its Implications for United States National Security*. Washington, DC: Environmental Media Services.

Scott, Anthony. 1974. "Fisheries, pollution, and Canadian-American transnational relations." *International Organization* 28 (4): 827–848.

Seliktar, Ofira. 2005. "Turning water into fire: The Jordan River as the hidden factor in the Six-Day War." *Middle East Review of International Affairs* 9 (2): 57–71.

Simon, Julian. 1981. *The Ultimate Resource*. Oxford: Martin Robertson.

Simon, Julian. 1989. "Lebensraum: Paradoxically, population growth may eventually end wars." *Journal of Conflict Resolution* 33 (1): 164–180.

Soroos, Marvin. 1997a. "The turbot war: Resolution of an international fishery dispute." In *Conflict and the Environment*, ed. Nils Petter Gleditsch, 235–252. Dordrecht: Kluwer Academic Publishers.

Soroos, Marvin. 1997b. *The Endangered Atmosphere: Preserving Global Commons*. Columbia: University of South Carolina Press.

Sprinz, Detlef, and Tapani Vaahtoranta. 1994. "The interest based explanation of international environmental policy." *International Organization* 48 (1): 77–105.

Sprout, Harold, and Margaret Sprout. 1962. *Foundations of International Politics*. Princeton, NJ: Van Nostrand.

Susskind, Lawrence. 1994. *Environmental Diplomacy: Negotiating More Effective Global Agreements*. Oxford: Oxford University Press.

Tir, Jaroslav, and Paul Diehl. 1998. Demographic pressure and interstate conflict: Linking population growth and density to militarized disputes and wars. *Journal of Peace Research* 35 (3): 319–339.

Ullman, Richard. 1983. "Redefining security." *International Security* 8 (1): 129–153.

Underdal, Arild. 2002a. "One question, two answers." In *Environmental Regime Effectiveness: Confronting Theory with Evidence*, ed. Edward Miles, Arild Underdal, Steinar Andersen, Jørgen Wettestad, Jon Birger Skjærseth, and Elaine Carlin, 3–45. Cambridge, MA: MIT Press.

Underdal, Arild. 2002b. "The outcomes of negotiation." In *International Negotiation: Analysis, Approaches, Issues*, ed. Viktor Kremenyuk, 110–125. 2nd ed. San Francisco: Jossey-Bass.

United Nations Treaty Collection. n.d. *League of Nations and United Nations Treaty Series Database*. Available at <http://treaties.un.org> (accessed May 13, 2010).

Weinthal, Erika. 2002. *State Making and Environmental Cooperation: Linking Domestic and International Politics in Central Asia*. Cambridge, MA: MIT Press.

Westing, Arthur, ed. 1986a. *Global Resources and International Conflict: Environmental Factors in Strategic Policy and Action.* Oxford: Oxford University Press.

Westing, Arthur. 1986b. "Appendix 2. Wars and skirmishes involving natural resources: A selection from the twentieth century." In *Global Resources and International Conflict: Environmental Factors in Strategic Policy and Action*, ed. Arthur Westing, 204–210. Oxford: Oxford University Press.

Westing, Arthur. 1986c. "Global resources and international conflict: an overview." In *Global Resources and International Conflict: Environmental Factors in Strategic Policy and Action*, ed. Arthur Westing, 3–20. Oxford: Oxford University Press.

Young, Oran. 1989. "The politics of international regime formation: Managing natural resources and the environment." *International Organization* 43 (3): 349–375.

Young, Oran. 1994. *International Governance: Protecting the Environment in a Stateless Society*. Ithaca, NY: Cornell University Press.

Zartman, William, and Jeffrey Rubin. 2000. "Symmetry and asymmetry in negotiation." In *Power and Negotiation*, ed. William Zartman and Jeffrey Rubin, 271–293. Ann Arbor: University of Michigan Press.

II

Scarcity and Degradation of Global Commons

2

Climate Change, Cooperation, and Resource Scarcity

Robert Mendelsohn

Climate change is certainly one of the most daunting environmental problems facing society because emissions come from all inhabited corners of the planet, the impacts affect the entire world, and the pollution is long lasting. The emissions of carbon dioxide, nitrogen dioxide, methane, and other greenhouse gases caused by human activities have led to a steady increase in the concentration of these gases in the atmosphere (IPCC 2007a). Activities as varied as growing rice, clearing forests, producing electricity, or driving a car all contribute to emissions. Because emissions of greenhouse gases stay in the atmosphere for long periods of time, the greenhouse gases accumulate in the atmosphere (IPCC 2007a). They then reflect infrared radiation (heat) back to the earth's surface, increasing surface temperatures. If this warming induces positive feedbacks, there can be large increases in surface temperatures. There has been a 0.5°C increase in temperature since the 1800s, including a 0.25°C increase in temperature since 1960. As the accumulation of greenhouse gases (a stock externality) increases, future temperatures are expected to climb, with increases of another 1° to 5°C by 2100 (IPCC 2007a). Although it is likely that the increase in temperature and gases experienced to date have been beneficial (Mendelsohn 2007), larger increases in temperature in the second half of this century will likely be harmful (Tol 2002; Mendelsohn and Williams 2007; IPCC 2007b). The marginal damages from greenhouse gas emissions will increase over time. Socially optimal abatement should therefore also increase over time (Nordhaus 1991).

In the next section, I present a bare-bones model of stock externalities that lays out the global warming problem. The model explains how the accumulating pollutant leads to rising marginal damages over time. I assume every country is alike. Each country generates an equal amount of pollution, and they are all affected equally by climate change.

The optimal cooperative solution to this problem is well known (Nordhaus 1991). Countries should equate the marginal cost of abatement, and this marginal cost should be equated to the marginal damages that emissions cause. I contrast this solution to a noncooperative outcome in which each country tries to "free ride," hoping that everyone else will cut back their emissions. This common property problem with greenhouse gases is also well known. Yet the problem is often cast in a static framework. This chapter emphasizes the dynamic nature of the greenhouse gas problem. As the resource (clean air) becomes scarcer, the problem worsens, pushing actors away from the noncooperative solution. Specifically, I argue that there is little at risk if the cooperative solution fails while climate warming causes little or no damage. If no one cooperates, there will be no abatement, but there will be few repercussions as well. As greenhouse gases accumulate and temperatures rise, however, the increasing damages associated with failing to control greenhouse gases will make the noncooperative solution increasingly unpalatable. There will be ever-mounting pressure on all the players (countries) to find a cooperative solution. The damages of unchecked global warming will eventually push countries to develop binding universal controls. At the same time, it is likely that such controls will not come into effect until the risks of climate change become more near term and threatening.

In the third section, the assumption that all countries are alike is relaxed, and I begin to explore the consequences of heterogeneity. What difference does it make if some countries grow more rapidly, have high carbon energy supplies, more expensive energy, are less developed, or will have different climate impacts? What will this do to their incentive to cooperate in an international treaty? I specifically discuss the Kyoto Protocol in light of these differences, and note why some countries participated but many did not.

In the final section, I use the lessons from each of these models to discuss how to design a more effective treaty in the future. How could the treaty be structured to get more countries to participate? How aggressive should the treaty be over time? What will it take to enforce the treaty?

A Stock Externality: Greenhouse Gases and Resource Scarcity

I start by looking at the simplest representation of the stock externality problem, and assume that all countries are identical. All countries face the same global stock externality. I wish to examine the incentives for countries to cooperate to control the externality versus to seek what is in their own individual best interests.

The following model provides a basic insight into how all stock externalities behave. Greenhouse gases are just a quintessential example of a pollutant that lasts over time (Nordhaus 1991). Although there are many studies that capture the behavior of stock externalities, I borrow heavily from Falk and Mendelsohn (1993) because of its simplicity.

I begin by examining the cooperative case. Every country cooperates, and the world effectively acts as a single agent. There are two costs: mitigation costs and climate damages. The objective is to minimize the sum of these two costs. The marginal cost of mitigation must be set equal to the marginal damage from climate change. Because the costs occur over time, though, the solution must account for the value of time. The value of time is solved daily for the entire world as investors and savers come together to determine a price for time: the interest rate. The "present value" of costs takes into account when the costs actually occur. Costs that are delayed have less of an impact. The optimal solution equates the marginal cost of mitigation with the present value of marginal climate damages.

Given that the present value of marginal damages is the same for every emission at a moment in time, the marginal cost of abatement should be the same for every source on the planet. Every sector, every firm, and every consumer should equate the marginal cost of abatement. The cooperative solution requires universal application.

Because the marginal damages of climate depend on the stock of greenhouse gases in the atmosphere, the marginal damages rise with time as the stock increases. Clean air becomes scarcer over time, and its marginal value rises. The optimal solution implies that the marginal cost of mitigation should rise over time with less and less emissions each year. The optimal solution should be dynamic. This general point is captured in figure 2.1. The optimal emissions should fall over time.

At first, the marginal damages are low. Equating the marginal damage to the marginal cost implies a large amount of emissions and little abatement. In this model, temperatures have risen by midcentury and so the marginal damages are higher. The higher marginal damages imply that there should be tighter regulations, higher marginal costs, and fewer emissions. By the end of the century, the marginal damages are much higher and regulations should have reduced emissions substantially. With the declining emissions each year, the cumulative emissions increase at a

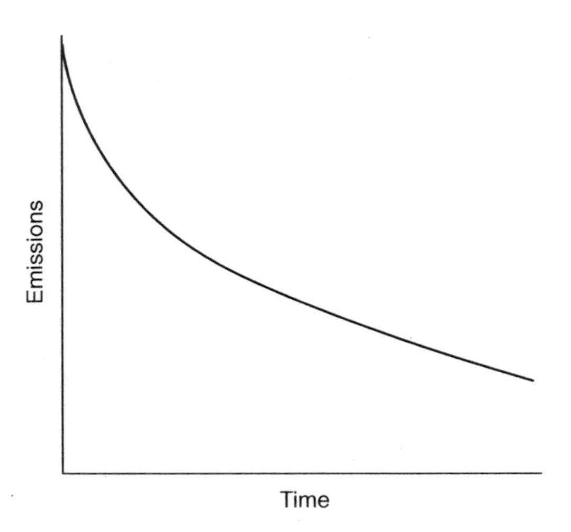

Figure 2.1
Optimal global emissions over time

slower rate over time. The cooperative solution over time is shown in figure 2.2. Both figures 2.1 and 2.2 are illustrative of the economics of greenhouse gases, and should not be interpreted literally. There is a great deal of debate about the exact marginal benefits (Pearce et al. 1996; Tol 2002) and costs of abatement (see, for example, Barrett 2003, 378). What all economists agree on, however, are the dynamics of the problem.

The marginal damages eventually rise to a point where emissions just offset the decay in the stock. At this point, the stock comes into equilibrium and does not change further. The damages no longer change, and emissions remain constant. The "optimal equilibrium" is an internal solution to this problem and not some arbitrary target that policymakers need to invent. The equilibrium level of the stock of greenhouse gases is uncertain, and a subject of active debate and research. Although carbon and methane cycles are understood at current levels of stock, it is not clear how they will behave as the stock increases. It is not clear what level of stock is the correct target and when we will reach it. The equilibrium level of greenhouse gases may, in fact, not be reached in this century.

The optimal dynamic solution described above is a cooperative one. It implies that all countries are working together to maximize the net benefits of the planet. Although the economics of climate change has been quick to identify this optimal solution, the policy community has found it hard to achieve. Why is it in practice hard to obtain this coop-

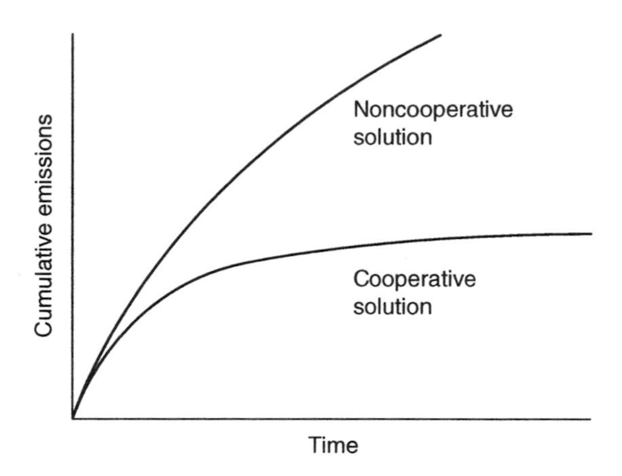

Figure 2.2
Cooperative and noncooperative cumulative emission paths for greenhouse gases

erative solution? Greenhouse gases are more challenging because of the global nature of the problem. Unlike most pollutants that are regional in nature, greenhouse gas emissions anywhere in the world contribute equally to the stock. The impacts of greenhouse gases are also global. If the climate across the entire planet changes, everyone is affected. Nations by themselves cannot affect the outcome. An effective control program must obtain global cooperation.

The noncooperative solution is for every country to examine the cost of its own contribution versus the benefit it will receive in return. If we assume that countries are small relative to the world, they alone will have little impact on the global levels of stock. The marginal benefit from mitigation will be exceedingly small. In contrast, every dollar they spend on mitigation is a dollar lost. The noncooperative solution, where only self-interest is part of the calculation, is to spend exceedingly little on mitigation. The problem is like the prisoner's dilemma with many players. As can be seen in figure 2.2, the noncooperative solution leads to higher cumulative emissions.

The global warming problem is not a one period game, however. Through regulations, countries decide how much to emit and abate every year. The game is replayed many times, and the outcomes of regulation choices change over time. As greenhouse gases accumulate (clean air becomes scarcer), the damages of emissions increase (see figure 2.3). In the early period, the marginal damages are low; in the middle period, the marginal damages rise; and in the late period, they are high. The marginal abatement cost function rises from right to left as more

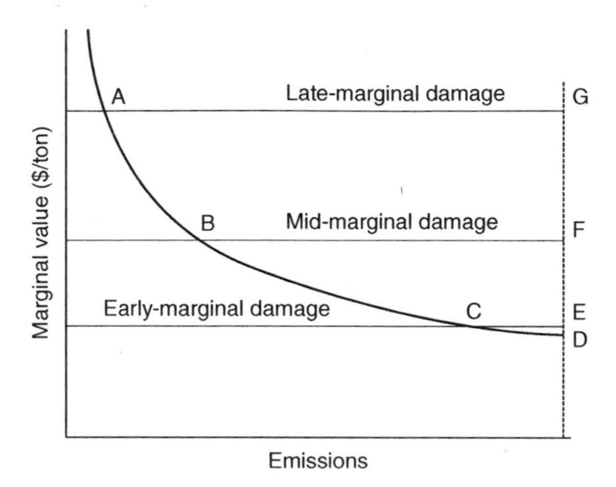

Figure 2.3
The net benefits of control as stock changes

emissions are reduced. At first, clean air is abundant, the damages from emissions are low, and the difference in the area between the marginal abatement cost function and the Early-Marginal Damage function is small (represented by the area CED). There is thus little reason to cooperate. If players do nothing, all they lose is this small net benefit. The failure to cooperate causes little anguish. Yet as greenhouse gases accumulate, the marginal damage of an emission increases (represented by the Mid-Marginal Damage function). The benefits of control remain the difference between the marginal damages and the marginal costs, but the marginal damages are much higher. The benefits of moving to a cooperative solution increase (they are now represented by the area BFD = BFCE + CED). In the late period, the benefits are even higher. The marginal damage is now the Late-Marginal Damage function. The gains from cooperation are represented by the area AGD = AGBF + BFCE + CED. The selfish incentives not to cooperate remain, but the downside of failing to reach a cooperative agreement starts to become untenable.

For example, the marginal damages per ton of carbon dioxide are likely to be in the range of US$0 to US$5 per ton at the moment. The difference between this benefit (from preventing the marginal damage) and the marginal abatement cost is small. Therefore, it does not really matter whether countries abate at the moment. By midcentury, the damages will probably rise to US$20 to US$50 per ton. The difference between the marginal benefits (from preventing these damages) and the marginal costs of abatement are larger. There is more incentive to find a cooperative solution to take advantage of these investments. By the end of the century, the marginal damages could well range from US$75 to US$200 per ton. The benefits of mitigation and preventing such damages will far outweigh the costs, and cooperation will become increasingly attractive. Scarcity thus increases the incentive to cooperate.

Although the marginal damage of carbon is uncertain, there is no question that the value of carbon will increase over time as the stock increases. The rising value of carbon provides increasing incentives to overcome national self-interest and find cooperative global solutions. Stock externalities give only small incentives to cooperate at first. Over time, though, the incentives escalate dramatically to the point where it is highly likely that countries will come to a binding agreement. In the case of global warming, the incentives may be small now. Countries may not agree to binding abatement programs in the near future. But as time passes, the problem will intensify, and it is ever more probable that a binding cooperative agreement will be reached.

In the simple scenario described above, all countries are assumed to be alike. They have symmetrical incentives to engage in greenhouse gas control. With all countries being alike, cooperative agreements can have simple rules that treat everyone alike. Relative to heterogeneous contexts, parties are less likely to require unique incentives or information (Compte and Jehiel 1997). The cooperative agreement can rely on uniform rules that apply to everyone equally. Presumably, the costs of negotiating an agreement in such a setting are relatively small.

Heterogeneous Countries

In this section, I explore what happens to this simple game if we assume the more realistic case: countries are not all alike. How will heterogeneity among countries affect greenhouse gas agreements? Will agreements be used to favor some countries over others? What insights do these perspectives provide concerning the Kyoto Protocol? I specifically examine five ways that countries vary from one another that are important to greenhouse gas policy: the rate of economic growth, the carbon content of energy supplies, the cost of energy, the level of development, and climate impact.

I assume that every country is going to look at whether they wish to join the agreement. They will do so only if they obtain a share of the net benefits of the social cooperative solution. Treaties that do not share the net benefits will not acquire universal participation. Without the near-universal participation of large emitters, global warming treaties are likely to be costly and ineffective.

The Rate of Growth

The average real rate of growth (adjusted for inflation) of the world GDP in 2005 was almost 5 percent (Central Intelligence Agency 2005). Yet the economies of individual countries are growing at widely varying rates. Because carbon emissions tend to be highly correlated with economic growth, countries that had high rates of growth also had large increases in their emissions. For example, China has been growing at 10 percent per year. Eastern Europe and the former Soviet Union have been growing at over 6 percent a year following the large recession at the break up of the Soviet Union. The United States has been growing at 3.2 percent annually, and Japan at 2.6 percent. Finally, other countries had relatively little growth, and so their emissions were relatively static. (Western Europe, for instance, grew at less than 2 percent a year.)

With the optimal cooperative solution, the marginal cost of abatement should be equated across the planet. If some countries have more rapid economic growth, they will need to increase their emissions over time relative to countries with slower economic growth. Otherwise, the fast-growing countries will face high marginal costs of mitigation relative to the marginal costs of the slow-growing countries. This will be wasteful, thereby easily doubling the overall cost of reaching any total mitigation target.

Unfortunately, the Kyoto agreement tied future emissions to historic levels. The agreement required greater reductions from rapidly growing countries. For example, U.S. emissions of carbon dioxide grew approximately 25 percent from 1990 to 2010, whereas European emissions (counting Eastern Europe) hardly grew at all. For the Europeans to reduce emissions to 7 percent below their 1990 level required only a 7 percent reduction. The United States had to reduce emissions by 32 percent to get back to 7 percent below its 1990 levels. The treaty placed a heavy burden on the United States, which is one of the reasons that the U.S. Senate rejected the Kyoto Protocol.

Of course, one way to make certain that the marginal cost of abatement is the same across the world is to permit countries to trade mitigation responsibilities (Victor 2004). Countries with low marginal costs can offer to buy permits from countries with high marginal costs. Open trading will establish a single world price for greenhouse gas emissions, thus assuring that the marginal costs across countries are equated. With trading, efficiency no longer depends on the initial allocation of permits. Trading assures that the permits are used where they are most needed. This works as long as every country is bound by the permit system. If some countries are bound by the trading agreement and others are not, trading will not necessarily get effective cooperation from the countries that are not included in the agreement. This is why it is crucial to get all large emitters into the agreement.

If the initial allocation is independent of the efficient solution, permits can be allocated to encourage countries to belong to the treaty. That is, the permit system can offer incentives for countries to join and enforce the treaty. The allocation can provide every country with a share of the net income of the cooperative solution. Trading thus frees the initial allocation of permits to be used as a tool to get countries to join the agreement.

The political economy problem associated with creating a successful agreement is that countries always want a bigger share of the global net benefits. But this is a zero-sum game. The global net benefit of controlling greenhouse gases perfectly efficiently is a fixed amount. If any one country gets more of these global benefits, other countries must give up some. A slow-growing country has an incentive to tie emissions to historic levels of emissions. This will force faster-growing countries to bear a bigger burden of the costs. Faster-growing countries want to tie permits to planned growth, and so they will automatically get more permits.

The slow-growing Western European countries and Japan successfully tied permits or targets at the Kyoto convention to 1990 emissions. Emission reductions were targeted relative to emissions on this particular date. The approach favored countries that subsequently grew little and especially those that went into recession. The approach hurt countries whose economies grew after 1990. Of course, when treaties are set to favor one group over another, they create a disincentive for the less-favored parties to participate. In this case, the parties that had no incentive to join the treaty were developing and rapidly growing countries such as the United States. It was no surprise that these parties did not agree to reduce emissions.

An alternative approach that would have been more advantageous to the United States and developing countries would have established a planned growth of emissions based on GDP growth. Each country would then have to reduce emissions relative to the planned growth by a constant percentage. The relative gain to the United States and developing countries certainly would be a relative loss to the slower-growing countries.

In this analysis, perfect cooperative solutions have been compared to noncooperative solutions where every country is on its own. In practice, one could also imagine smaller coalitions of countries forming an agreement. If the omitted nations generated little greenhouse gases, it would not matter. For example, it does not matter that many poor developing countries are included because they emit so few greenhouse gases. It does matter that emerging nations (e.g., China, India, Mexico, and Brazil) and the Organization for Economic Cooperation and Development (OECD) are included because they are responsible for the bulk of emissions. Coalitions of only some of these countries, however, are likely to be ineffective. If the small coalitions spend vast sums on abatement while the rest of the world spends nothing, the program will cost the coalition dearly and yet have little effect on the global outcome (Nordhaus 2008).

The Carbon Content of Fuels

Another important difference among countries is that some have low carbon energy supplies whereas others have abundant energy supplies that have a high carbon content. Specifically, some countries have large amounts of renewable energy in the form of hydropower, geothermal, or biofuels. As carbon prices rise, the relative price of renewable energy will rise, making these resources far more valuable. Even fossil fuels that are low in carbon such as natural gas will become more valuable. In contrast, with regulation, the price of fossil fuels high in carbon will fall. Countries with large supplies of oil, shale oil, and especially coal will see a dramatic reduction in their wealth. With severe regulations and no new technology, they may well find that they cannot even extract the resource. The cost of carbon regulations will consequently fall heavily on countries with substantial coal deposits such as the United States, China, Canada, Australia, and Germany.

The Cost of Energy

Another significant difference across countries concerns the cost of energy. Although international trade should equate the cost of energy from one country to the next, some countries have adopted relatively extreme energy policies that make energy expensive within their borders, and other countries subsidize it. For example, Europe places high taxes on imported oil, making oil and diesel prices in the European Union (EU) much higher than the rest of the world. The average price of gasoline in the EU is about US$5.50 compared to US$2.50 in the United States.

Taxing carbon emissions in high-energy-cost countries is especially expensive because one is adding the carbon tax to a set of energy taxes that are already very high. The welfare cost to these countries of a carbon tax is higher than to other countries. High-energy-cost countries should consequently be reluctant to impose carbon regulations. If high-energy-cost nations can use the carbon regulations to make energy expensive in other countries, they can gain a competitive advantage. That is, the treaty would be attractive to high-energy-cost countries if they can use the carbon regulations to entice their competitors to also adopt policies that make energy expensive. This would make their energy-efficient products and technology suddenly attractive to the rest of the world. A carbon program can be used as a tool to make low-energy-cost countries adopt policies that make energy expensive. The high-energy-cost countries were especially eager that the United States adopt policies that would make U.S. energy expensive. Of course, once again, this limited the incentive for the United States and other low-energy-cost countries to participate. Unfortunately, if the goal of the policy is to make energy expensive in a foreign country, the treaty might not care if the policy is also inefficient. The original Kyoto negotiators wanted to ban the use of trading, for example, thereby forcing energy prices to rise in the United States. It was only after the United States dropped out of the agreement that the Kyoto negotiators accepted trading as a suitable tool.

Another problem that the Kyoto negotiators faced was the "hot air" caused by the collapse of the Soviet economy. As the economies in the former Soviet Union disintegrated, their emissions fell below their 1990 historic levels. The countries had negative commitments according to the rules of the Kyoto agreement. The former Soviet bloc technically could sell these commitments without having to do any mitigation. Of course, this did not result in any actual reduction in carbon emissions and so was labeled hot air. Negotiators interested in earnest reductions in emissions did not want countries merely trading with the former Soviet Union countries for hot air permits. Still, this hot air has allowed the EU to appear to have reduced emissions substantially without really spending much on mitigation.

Developed versus Developing Countries

Another source of heterogeneity concerns development. If one looks at the past, the OECD countries have been responsible for 75 percent of the cumulative emissions (IPCC 2007a). At this point, however, the OECD countries are responsible for less than half of all emissions, and that fraction is falling all the time (IPCC 2007a). The emerging countries (such as China, India, and Brazil) already emit as much carbon as the advanced economies. Should permits be allocated on the basis of the cumulative historic pollution or the current pollution? Should the advanced economies be solely responsible for abating greenhouse gases, or should the emerging countries share a role? If controlling greenhouse gases limits growth, should the emerging countries be asked to pay only when they enjoy comparative levels of income per capita?

Once again, the issue is how to allocate the initial set of tradable permits. One could distribute them in proportion to the global GDP. For example, to keep emissions from growing, each country would be allowed to emit a ton of carbon dioxide per US$2,700 of GDP (or US$60 trillion GDP per 22 billion tons). Clearly this would grossly favor the OECD countries, which currently have a large share of the world's GDP. Alternatively, emissions could be given out per person. Each country would be given 3.5 tons of carbon dioxide per person (or 22 billion tons per 6.3 billion people). This would be much tougher on OECD countries, which have only about 16 percent of the world's population. The OECD countries would have to buy permits from the developing countries. Clearly, the OECD counties would be more eager to create a carbon per GDP plan, and the developing countries a carbon per capita plan.

The Kyoto Protocol was based on actual emissions so that it more closely resembled the carbon per GDP plan. It is no wonder that developing countries expressed little interest in joining the agreement. In order to balance the interests of developing and developed countries, one would need a compromise between the two extremes suggested above.

Impacts

The original conception of global warming was that everyone was going to lose because the climate was changing. That is, the current climate was assumed to be optimal for every country so that any changes would be bad for everyone. Most of the negotiators involved with the Kyoto Protocol were focused on the costs of mitigation, not the damages from impacts. Yet subsequent research has shown that the impacts of climate change will not be evenly distributed across the planet. For the next fifty years, mid- and high-latitude countries are likely to benefit from warming, whereas the damages from warming will probably be concentrated in the low latitudes (Kurukulasuriya et al. 2006; Mendelsohn and Williams

2004; Mendelsohn, Dinar, and Williams 2006; Tol 2002). Mid- and high-latitude countries are expected to receive large gains in the agricultural sector. A warmer, wetter, carbon dioxide–enriched planet will be especially kind to the mid and high latitudes. In contrast, many developing countries are already located in places that are too hot or dry. Further warming will exacerbate the climate constraints that farmers in these countries already face. Climate change throughout the century is likely to have a disproportionate impact on low-latitude countries (Mendelsohn, Dinar, and Williams 2006).

The likelihood that the bulk of the damages from global warming will fall on poor low-latitude countries creates additional problems for a global treaty on greenhouse gases. Many countries that currently emit a lot of carbon actually will benefit from doing little abatement, at least for the next half century. The countries that are hurt the most, not to mention first, by climate change are the poor low-latitude nations that emit little carbon.

In principle, poor low-latitude countries could pay more wealthy mid- and high-latitude countries to reduce their carbon emissions. The equity problem with this and the imbalance of power makes it an unlikely solution, however. Alternatively, the wealthy mid- and high-latitude countries could accept responsibility for their emissions, and fund the bulk of carbon control. Will nations take on this responsibility once they understand that the primary benefits go to poor people in the low latitudes? Will nations accept the responsibility for their actions? Would mid- and high-latitude countries curb their emissions if they understood how their actions impacted the poor low-latitude countries of the world?

Conclusion

The optimal cooperative solution to global warming involves a dynamic set of regulations that tighten over time. As the resource (clean air) becomes scarcer, regulations should tighten. The marginal damage of a ton of emitted carbon is initially low, but it rises gradually as the stock of greenhouse gases accumulates. The marginal cost of abatement should consequently be relatively low at first and then rise as the present value of the marginal damages increases. This optimal solution yields the highest total net benefits for the world from cooperating to control greenhouse gases.

Optimal cooperation is nevertheless difficult to achieve in the real world with so many actors having their own incentives to opt out.

Because the gains from cooperation are initially quite small, it should perhaps not be a surprise that an effective international agreement has not yet been crafted. The net gain from a global warming treaty at the moment is small. While the resource (clean air) is abundant, the cooperative solution provides few incentives. The difficulty of achieving an international agreement is not sufficiently strong to overcome all the hurdles.

As greenhouse gases accumulate and clean air becomes scarce, the incentive to cooperate will increase. Warming becomes ever-more harmful, and the gain from cooperative mitigation increases. Countries have an ever-greater incentive to enter into a cooperative agreement. The growing scarcity makes the cooperative solution ever-more attractive. Analysts should not infer that the current failure of the Kyoto agreement necessarily means that future agreements will also fail.

In fact, the Kyoto negotiations led to some experiences that may help the world eventually craft a more effective treaty. First, global warming treaties must be efficient. Efforts to make the treaty wasteful (inefficient) rob the treaty of its potential net benefits. Without enough global net benefits, there will be no incentive for countries to join or behave. The treaty must equate the marginal cost of abatement across sources from all countries in each time period. For example, the treaty should encourage permit trading as much as possible. The treaty should include carbon sequestration as well as mitigation. The treaty must also choose efficient targets over time. This is a dynamic problem. Too much abatement too early, or else too little, too late, reduces the effectiveness of the program. Timing is important. The treaty must reflect how scarce the resource has become.

Second, the treaty must share the net benefits of cooperation across member states so that each state has an incentive to remain in the treaty. Efforts to use the treaty as a tool to gain advantage over other countries will lead to problems with participation and enforcement. Countries that are harmed by the treaty will work to undercut it. Attempts to load the cost of abatement on to countries that do not volunteer to take on this burden will lead to resentment and poor performance.

Third, the discrepancy between who emits greenhouse gases and who suffers the damages of climate change creates a special problem for climate change treaties. Although greenhouse gases effectively make clean air scarce, there are much more severe consequences for low-latitude countries than for the rest of the world. Global warming is an urgent matter for the low-latitude countries. Low-latitude countries view this "scarcity" differently from mid- to high-latitude countries. The mid- and high-latitude countries will not experience the same consequences from warming. They will not view global warming as urgent because they have less at stake. Unless all emitting countries take on the responsibility of the consequences of their actions, even if those consequences occur in distant places, there is little chance of crafting a successful treaty. It is not likely that the poor countries of the low latitudes can pay the rich mid- and high-latitude countries to abate. A successful arrangement will depend on the emitting countries agreeing to their liability for what happens to the poor in the low latitudes.

Although the Kyoto Protocol has proven to be an ineffective treaty for greenhouse gas controls, a well-crafted future agreement could well succeed. First, as the damages from climate change rise, the incentives to engage in binding international negotiations increase. That is, as the resource "clean air" becomes scarce, society will want to take measures to preserve what is left in the future. Second, there is time to develop an efficient treaty. It is more important to produce a long-lasting agreement than to rush into a poorly constructed one. The net benefits from controlling greenhouse gases are small because the costs of control are high. A treaty will not be long lasting if the net benefits are squandered by inefficient design. Third, a well-crafted treaty will try to share the net benefits across countries so that they participate without undue enforcement costs. Rather than using the treaty as a tool to punish countries or gain advantage, a well-designed treaty would try to create incentives for every country to want to join and enforce the treaty. Yet for all of this to work, emitting countries must take responsibility for the damages that their emissions cause to others. The rich mid- and high-latitude countries must agree to their liability to the poor low-latitude countries that will suffer the bulk of the damages from climate change. When these conditions are met, there is every reason to believe that a successful climate change treaty will be created and enforced.

In the meantime, international negotiators should seek lesser targets. For example, it would be a helpful start simply to measure emissions from every source. Although this simple reporting task has no binding regulations, it has been shown in the past to encourage firms to reduce emissions. Further, knowing where the sources are is a crucial first step in regulation. Individual countries should explore alternative policy mechanisms to control greenhouse gases. The regulations do not have to be expensive. They merely demonstrate that a country is taking action.

By experimenting with alternative regulatory methods, countries can learn how best to approach more vigorous regulations when needed. Countries can begin to explore technologies to remove carbon and generate noncarbon energy. By looking into alternative technologies, it might be possible to identify low-cost strategies to control greenhouse gases that are not evident at the moment.

References

Barrett, Scott. 2003. *Environment and Statecraft: The Strategy of Environmental Treaty-Making*. Oxford: Oxford University Press.

Central Intelligence Agency of the United States. 2005. *The World Factbook*. Falls Church, VA: Central Intelligence Agency.

Compte, Olivier, and Philippe Jehiel. 1997. "International negotiations and dispute resolution mechanisms: The case of environmental negotiations." In *International Environmental Negotiations: Strategic Policy Issues*, ed. Carlo Carraro, 56–70. Cheltenham, UK: Edward Elgar.

Falk, Ita, and Robert Mendelsohn. 1993. "The economics of controlling stock pollution: An efficient strategy for greenhouse gases." *Journal of Environmental Economics and Management* 25 (1): 76–88.

IPCC (Intergovernmental Panel on Climate Change). 2007a. *The Scientific Basis*. Cambridge: Cambridge University Press.

IPCC (Intergovernmental Panel on Climate Change). 2007b. *Impacts, Adaptation, and Vulnerability*. Cambridge: Cambridge University Press.

Kurukulasuriya, Pradeep, Robert Mendelsohn, Rashid Hassan, James Benhin, Temesgen Deressa, Mbaye Diop, Helmy Mohamed Eid, K. Yerfi Fosu, Glwadys Gbetibouo, Suman Jain, Ali Mahamadou, Renneth Mano, Jane Kabubo-Mariara, Samia El-Marsafawy, Ernest Molua, Samiha Ouda, Mathieu Ouedraogo, Isidor Se´ne, David Maddison, S. Niggol Seo, and Ariel Dinar. 2006. "Will African agriculture survive climate change?" *World Bank Economic Review* 20 (3): 367–388.

Mendelsohn, Robert. 2007. "Past climate change impacts on agriculture." In *Handbook of Agricultural Economics*, ed. Robert Evenson and Prabhu Pingali, 3: 3009–3031. Amsterdam: North-Holland.

Mendelsohn, Robert, Ariel Dinar, and Larry Williams. 2006. "The distributional impact of climate change on rich and poor countries." *Environment and Development Economics* 11 (2): 159–178.

Mendelsohn, Robert, and Larry Williams. 2004. "Comparing forecasts of the global impacts of climate change." *Mitigation and Adaptation Strategies for Global Change* 9 (4): 315–333.

Mendelsohn, Robert, and Larry Williams. 2007. "Dynamic forecasts of the sectoral impacts of climate change." In *Human-Induced Climate Change: An Interdisciplinary Assessment*, ed. Michael Schlesinger, Haroon Kheshgi, Joel

Smith, Francisco de la Chesnaye, John Reilly, Tom Wilson, and Charles Kolstad, 107–118. Cambridge: Cambridge University Press.

Nordhaus, William. 1991. "To slow or not to slow: The economics of the greenhouse effect." *Economic Journal* 101 (407): 920–937.

Nordhaus, William. 2008. *A Question of Balance: Weighing the Options on Global Warming Policies*. New Haven, CT: Yale University Press.

Pearce, David, William Cline, Amrita Achanta, Samuel Fankhauser, Robert Pachauri, Richard Tol, and Paul Vellinga. 1996. "The social cost of climate change: Greenhouse damage and the benefits of control." In *Climate Change 1995: Economic and Social Dimensions of Climate Change*, ed. James Bruce, Hoesung Lee, and Erik Haites, 183–224. Cambridge: Cambridge University Press..

Tol, Richard. 2002. "Estimates of the damage costs of climate change. Part 1: Benchmark estimates." *Environmental and Resource Economics* 21 (1): 47–73.

Victor, David. 2004. *The Collapse of the Kyoto Protocol and the Struggle to Slow Global Warming*. Princeton, NJ: Princeton University Press.

3

Ozone Depletion: International Cooperation over a Degraded Resource

Elizabeth R. DeSombre

International cooperation to protect the ozone layer is frequently seen as one of the hallmarks of successful international environmental cooperation. In response to potential future harm to a shared resource, states tried to limit their use of important industrial chemicals—before the extent or causes of the problem were fully understood—and agreed to precedent-setting measures to help developing countries adapt to development without these substances. Although there have been problems with black markets in controlled substances (Clapp 1997) and some degree of noncompliance from states with the timing of emissions reductions, these issues have ultimately been minor (DeSombre 2001, 74), and have been better managed and understood than in almost any other system of international environmental cooperation. The international cooperation to address ozone depletion deserves the positive reputation it has achieved.

Scarcity and environmental degradation certainly played a role in promoting international cooperation, but other factors and related bargaining tactics were likewise fundamental to understanding this cooperative exercise.

Ozone, Conflict, and Cooperation: Scarcity, Degradation, and Bargaining Power

Why examine this environmental problem in the context of a volume on conflict and cooperation under conditions of resource scarcity and degradation? Although potentially contentious, the successful cooperation that ozone depletion represents makes it an unusual case for a collective consideration of scarcity and conflict. In addition, the ozone layer is not the type of resource one usually thinks of when contemplating the scarcity or overuse of resources. States "use" the ozone layer to provide protection—a use that does not in itself diminish the amount of ozone for others to take advantage of, unlike the use of fishery resources or clean water. Moreover, to the extent that the resource is degraded (or the protections given by the ozone layer become scarcer), any specific degradation can only be traced indirectly to direct actions by people in specific states, even if human responsibility for the degradation broadly is well understood. The environmental problem is diffuse and long term in a way that makes assessing specific responsibility conceptually much less clear than in most cases of scarcity or degradation. So although there are impacts on the resource by actions of individuals and states, these effects do not follow the traditional patterns of scarcity and degradation that we are most familiar with, and consequentially they have different ramifications for how the politics of these issues are negotiated.

This case is nevertheless instructive for a broader consideration of the relationship between scarcity and degradation, on the one hand, and environmental cooperation and conflict, on the other. Environmental degradation of a global commons resource generally requires international cooperation to mitigate. Because the ozone layer is a shared resource, outside the legal purview of any individual state and affected by actions taken within any of them, it is an issue that cannot be addressed successfully without international action. It is a nonexcludable resource, meaning that actors cannot be prevented (either legally without their own consent or practically) from having an effect on the resource. Likewise, the resource is rival: actions taken by one actor can diminish the usefulness of the resource for others (Barkin and Shambaugh 1999, 3). Cooperation is therefore necessary to convince all relevant parties to mitigate their contribution to the degradation. Without the threat of degradation of the resource, there would be no need for cooperation.

Ozone depletion is also not an issue that any state could address alone. Although the United States took domestic action before international regulation was considered, banning the use of some ozone-harming substances in nonessential aerosols in 1977 (Clark and Dickson 2001, 266), no one seriously considered acting alone as a reasonable alternative to international cooperation. Conversely, some developing states indicated a clear willingness to unilaterally (or collectively) continue or increase their ozone-depleting activities if their concerns were not addressed within the agreement. These threats gave developing states bargaining power. Even though these states would be harmed by ozone depletion, their willingness to prioritize their economic development over environmental action meant that any other states that wanted to address the problem collectively had to entice developing states to participate.

Moreover, with uncertainty about the extent of ozone depletion (and its causes) prominent, an increase in the perceptions of degradation led to an increase in the willingness to act. Uncertainty is a feature of most environmental issues: How many fish can be sustainably caught? How much rain will fall in this area next year? What kinds of local impacts will global climate change deliver, and when? Resolving uncertainty is not necessarily a panacea, but the case of ozone depletion illustrates that an increasing belief about potential—albeit still uncertain—environmental impacts can increase the likelihood of action. At the same time, however, cooperation began before there was a clear understanding of the extent of environmental degradation, so this case also illustrates that absolute evidence of environmental degradation is not necessary for cooperation and that uncertainty does not preclude action.

Other characteristics of the ozone depletion issue contribute to our understanding of how degradation can relate to cooperation, and specifically, how conflictual—and in what ways—those efforts at cooperation will be. Although the cooperative efforts to address this environmental problem are justifiably lauded, the ease of coming to this cooperation should not be overstated, and the outcome was by no means preordained. In particular, a variety of conflicts threatened to undermine the collective ability to mitigate the problem collectively, and examining them can help illuminate the intersection between conflict and cooperation when addressing resource issues.

Ozone depletion features some of the characteristics of issues considered in this volume, in ways that can help a broader consideration of the role of these phenomena. It features asymmetry in terms of both the socioeconomic conditions of the countries most involved in addressing the problem and the effects of environmental degradation. These asymmetries lead to some of the important dynamics in the negotiation over how to address the issue, sometimes in counterintuitive ways. At a minimum, they suggest that negotiating power in international environmental agreements may come from factors other than the types of traditional measurements of power that most scholars of international relations examine. Other important asymmetries include geography and time.

There are thus a number of lessons from this case for a broader consideration of the relationship between environmental degradation and international cooperation. Mitigating the degradation of international common pool resources requires cooperation. The extent of the cooperation can be increased by evidence of the seriousness of the degradation, but is not dependent on such evidence. In this context, in which a rival resource necessitates the involvement in its protection of all who could potentially harm it in the future, bargaining power comes to those who can credibly claim to care less about the resource. In international environmental politics that group of actors may include developing states (whose disinterest in addressing the environmental issue in question may come from prioritizing economic development), giving them more bargaining power than they have in most other international issues. Despite all these difficulties, the efforts at international cooperation to address ozone depletion show that such cooperation is possible, and that, even under difficult circumstances, international environmental agreements can have dramatically positive behavioral and environmental outcomes.

The Nature of the Problem

It is useful to begin with some brief background on the problem and the history of ozone depletion in order to understand the nature of the environmental issue as well as the subsequent political process of cooperation to address it. The ozone layer is a layer of the molecule O_3 approximately 12 to 25 kilometers above the earth. While ozone is a pollutant at ground level, in the stratosphere it performs essential functions, including restricting the amount of ultraviolet-B radiation that can reach the earth and therefore protecting life from receiving dangerous amounts of this radiation. Ozone in the stratosphere is naturally created and destroyed, and without human interference exists roughly in equilibrium (Parson 2003, 3–4).

The awareness of potential threats to the ozone layer came about in what was at that point a nontraditional manner of scientific understanding about environmental problems. Rather than noting effects and seeking to determine the cause, in this case the potential for an environmental problem from specified causes was predicted before the effects were evident. Beginning in the 1970s, scientists hypothesized that chlorofluorocarbons (CFCs)—a set of safe, nontoxic, and stable industrial chemicals—could persist in the atmosphere and travel to the ozone layer. There, in the presence of solar radiation, they could break down and cause a chain reaction that could destroy thousands of ozone molecules (Molina and Rowland 1974, 810–812). The scientific research to evaluate this hypothesis had not been fully resolved when international negotiations started to address this potential problem.

The nature of the ozone layer makes it an unusual "resource." Unlike water, fish, or trees, which can be overused or overharvested, when scholars refer to ozone "depletion," this depletion does not come from using too much of the ozone layer. In its use, the ozone layer functions much more like a public good than like a common pool resource: it is available to everyone, and the use by anyone of the shield it provides from ultraviolet (UV) radiation does not diminish anyone else's ability to use it for these purposes. It does share important characteristics with a common pool resource, however, when you consider it to be a "sink" for pollution. Some view the common pool resource as the provision rather than the use of a healthy ozone layer (Downie 1999, 102). Its ability to continue to provide UV protection is the rival aspect; this ability can be diminished by human activity (the emission of ozone-depleting substances). If there is a finite amount of ozone-depleting substances that can be used before the ozone layer is degraded, the use of these substances by one actor reduces the amount that can be used by others without causing damage to the ozone layer. Ozone depletion therefore involves both the rivalness and nonexcludability of other international environmental problems, even if those characteristics might not be initially obvious. Both of these aspects mean that the problem cannot be addressed without widespread, collective action.

The Montreal Protocol

Efforts to address the degradation of the ozone layer began with scientific cooperation in the late 1970s. The United Nations Environment Programme (UNEP) brought together scientific experts from thirty-two states in 1977 to assess the state of scientific knowledge; this conference produced the World Plan of Action and a coordinating committee to continue international efforts to address ozone depletion. In 1981, UNEP authorized negotiations toward an international agreement, eventually resulting in the Vienna Convention on the Protection of the Ozone Layer in 1985 (Chasek 2001, 103–104).

Disagreement among states about the seriousness of the potential environmental problem and the extent to which emissions reductions should be undertaken meant that the Vienna Convention emerged as a framework convention, laying out a set of general principles and a process for further cooperation, but that the agreement included no binding abatement obligations. The European Union (EU) and Japan, which had not yet undertaken domestic measures to address ozone depletion, resisted efforts by states—such as the United States, Canada, Australia, and the Nordic countries—to impose binding international emission rules (DeSombre 2006, 108). Instead, states agreed on the general obligation to "protect human health and the environment against adverse effects . . . from human activities which modify or are likely to modify the ozone layer" (Vienna Convention 1985, article 2 [1]). In addition, they agreed to cooperate on research about the problem of ozone depletion and harmonize policies to reduce human activities likely to negatively impact the ozone layer (Vienna Convention 1985, article 2 [2]). Part of the compromise in leaving specific emissions reductions out of the obligations in the Vienna Convention was an agreement to convene a working group to begin negotiation toward a protocol (Chasek 2001, 104).

Negotiations toward what became the Montreal Protocol began in 1986 and concluded a year later. There were serious disagreements over the form (whether to control consumption or production) and depth (how much of a reduction there should be, and of what and by when) of abatement measures. Ultimately the negotiating states agreed that developed countries would freeze their consumption—defined as "production plus imports minus exports" (Montreal Protocol 1987, article 1)—of the major CFCs at 1986 levels by 1990 and then reduce consumption in stages to 50 percent by 1999. Halon consumption, similarly, would be frozen at 1986 levels by 1992 (Parson 2003, 240–241). Developing states with less than 0.3 kilograms per capita of the regulated substances would be granted an additional ten years before having to meet these obligations (Montreal Protocol 1987, article 5). States within the agreement agreed to trade controlled substances (during the times that they were still legal to use) only with other states within the agreement. In addition, in the initial version of the agreement, developing states were offered vague promises of technical and financial assistance in meeting the protocol's obligations. As discussed below, threats by the major developing states to remain outside the agreement without guaranteed financial assistance to pay the costs of meeting the treaty's obligations resulted in the creation of the Montreal Protocol Multilateral Fund to transfer funding for the phase out of ozone-depleting substances from developed to developing states.

The results of the Montreal Protocol have been dramatic. With small exceptions for essential use, developed states have entirely phased out their use of the major CFCs, halons, carbon tetrachloride, and methyl chloroform, and they accomplished this phase out on an accelerated schedule. They have reduced their use of the additional regulated substances. Developing states for the most part have followed their reduction obligations as well, when these obligations have come due. And the environmental impact of these changes is evident. Concentrations of ozone-depleting substances in the atmosphere are no longer increasing and are even beginning to decline (Ozone Secretariat n.d.). Although that circuitous phrasing makes the atmospheric results sound modest, the long lifetimes of substances means that some of the first CFCs emitted in the late 1920s, when they were first invented, still persist in the atmosphere. A reduction in the increase of these substances and the start of a decline are impressive results indeed. Likewise, since the presence of ozone-depleting substances in the stratosphere has only just peaked, improvements in the ozone layer itself should not yet be expected. Current estimates suggest that the ozone layer over the Antarctic will recover by 2065 (British Antarctic Survey 2005). Less vulnerable areas of the ozone layer, also harmed by human activity, should also be expected to recover by that date.

The Many Asymmetries of Ozone Depletion

Cause and Effect

States are unequal in the extent to which they cause or are likely to suffer the consequences of ozone depletion, and these asymmetries have had an important effect on the efforts to protect the resource. In terms of cause, ozone-depleting substances (CFCs and halons, as the initial substances of concern, with hydrochlorofluorocarbons, hydrobromofluorocarbons, and methyl bromide later determined to cause problems as well) are primarily used in industrialization, for such activities as refrigeration, air-conditioning, fire suppression, and as blowing and cleaning agents. (Even methyl bromide, an agriculture fumigant, is used in industrial rather than subsistence agriculture.) Developed countries were thus primarily responsible for the manufacture and use of these substances. Developing countries, however, would have begun to produce and use ozone-depleting substances to an increasing extent in the future (Friends of the Earth 1990). The production of CFCs is relatively inexpensive and technologically simple, making their increased use by developing countries particularly likely, absent any prohibition of such use.

Geography

Ozone depletion is a standard common pool resource issue. The location of the harm caused is not related in any way to the location of emissions. That means that those states that emit the greatest amount of ozone-depleting substances are not necessarily those that will suffer the greatest consequences. On issues such as climate change, these discrepancies can be systematically unfair, with the problems created by developed states most acutely felt by developing ones. The ozone-depletion situation is more complicated, in a way that illuminates important aspects of North-South relations, but that also makes some aspects of addressing the problem easier.

There are a number of factors that influence vulnerability to ozone depletion. In terms of geography, the greatest depletion in the ozone layer happens over the poles, because the polar stratospheric clouds that form in this region provide a particularly hospitable location for the reactions that destroy the ozone (Parson 2003, 147–153). It is for these reasons that the "ozone hole" (more accurately a dramatic thinning) happens over Antarctica, where the ozone has decreased during these seasonal thinnings to only one-third of its pre-1975 levels (NASA Advanced Supercomputing 2001). A similar but slightly less dramatic thinning exists over the arctic. The worst effects from ozone depletion can be found near these locations, which affects developed countries in the northern hemisphere, and a mix of developed (e.g., Australia) and middle-income developing (e.g., Chile and Argentina) states in the southern hemisphere.

Another measurement of vulnerability is the extent to which humans feel the actual effects of this depleted ozone layer, resulting in increasing skin cancer rates or immune deficiency. Detlef Sprinz and Tapani Vaahtoranta (1994) use this approach, evaluating the extent of skin cancer cases as a way to assess environmental vulnerability to ozone depletion. By this measure developed countries, at least initially, seem the most vulnerable. Other ecosystemic effects of particular human concern—since plants can be vulnerable to the same cell destruction from increased UV radiation—include crop damage (which could be particularly harmful to those especially dependent on agriculture) and the potential decrease in ocean productivity that would harm those states most dependent on fishing. While the ecosystemic effects may be more problematic for some developing countries for whom these resources are a greater part of their overall economies, developing states are not systematically more vulnerable to ozone depletion than are developed ones. Some states in either category are more vulnerable than others in either category.

Interestingly, concern about the problem did not necessarily match the actual vulnerability. At the time of initial cooperation, it was not yet clear that the ozone layer *was* depleted and the effects had not yet been measured (including the potential human problems that might result both probabilistically and long after exposure), and thus it is understandable how concern about the problem could become disentangled from the empirical evidence about greatest harm. No one actually knew who was most vulnerable, so that subset of actors was unable to express the greatest concern. In the case of ozone depletion, it was developed countries in the Northern Hemisphere generally that were the most concerned about the problem. Public opinion in the United States especially, but also in Europe, favored action to protect the ozone layer as early as the 1970s, when the work of Mario Molina and Sherwood Rowland first began to attract attention (Clark and Dickson 2001, 265–267). This concern led the leadership in these developed states to take unilateral domestic action and push for international negotiations, even though there was no evidence that they were any more vulnerable to ozone depletion than any other states.

Socioeconomic Differences

Most important, developing countries maintained their disinterest in addressing the issue, especially compared to their other priorities. To some extent this disinterest was probably genuine. Compared to many of the economic and local environmental problems facing poor countries, the potential for future increases in skin cancer and decreases in agricultural productivity may have seemed remote. (Some even argued that darker-skinned people would be less vulnerable to skin cancer, although other immune system responses to increased UV radiation would likely affect people despite their melanin levels.) More important, the substances that would likely be regulated had been crucial to developed countries in their process of industrialization, and developing countries were justifiably suspicious of being told that they could not use the same cheap, nontoxic, stable substances that the North had used to industrialize.

But this disinterest was also politically useful and served to give developing countries the unusual degree of bargaining power that they had in negotiations to address ozone depletion. The rival nature of ozone depletion offered power to those who contributed to the problem yet didn't have a strong interest in addressing it. Without the participation of rapidly industrializing developing countries in measures to protect the ozone layer, any action undertaken by states that agreed to reduce their emissions could eventually be undone by the increasing use of ozone-depleting substances by these developing states. Developing countries had a credible threat to remain outside this process: they were not overly concerned about the environmental problem in question, and *were* concerned about the economic and development implications of undertaking action to protect the ozone layer.

The initial negotiations that produced the Montreal Protocol showed an awareness of the need to entice developing states into the agreement, but the mechanisms created—allowing a ten-year time lag before the developing states would have to meet the agreement's obligations, and restricting trade in ozone-depleting substances to states within the agreement—failed to persuade these states to join. The trade restrictions would only work if all the states capable of producing ozone-depleting substances ratified the agreement. But only one of the developing country producers at the time—Mexico—ratified the agreement immediately. And even among those not yet producing CFCs, doing so was a fairly simple process, so others could start if their access to these chemicals was restricted by those within the agreement. As long as key developing country producers—or potential producers—of ozone-depleting substances remained outside the Montreal Protocol, the agreement's trade restrictions would not work to force the developing country consumers of these substances to join.

Immediately after the Montreal Protocol negotiations were complete, only three developing states (Mexico, Nigeria, and Venezuela) signed and ratified the agreement. The others, led by China and India, indicated that until a mechanism for transferring funding from developed to developing states to meet Montreal Protocol requirements was specified, they would remain outside the process (Benedick 1991). Malaysia characterized the protocol as "inequitable" and argued that developing states did not realize its full implications for them during the initial negotiations (Friends of the Earth 1990). With the predicted growth in the use of ozone-depleting substances by these countries, the entire cooperative enterprise could have been undermined by their nonparticipation. They therefore had a great deal of bargaining power.

With this bargaining power, developing countries gained the creation of the Montreal Protocol Multilateral Fund, an institution in which developed states contribute funding that then goes to meet the "full incremental costs" for developing states of implementing the agreement's obligations. Not only was the elaboration of such a fund (when the Montreal Protocol itself only contained vague promises of financial assistance) significant but the processes to be followed for the allocation of funding were also notable and due to the insistent negotiating on the part of developing states (DeSombre and Kauffman 1996). Unlike existing organizations that allocate decision-making authority based on the extent of the contributed funding, the fund gives largely equal decision-making power to developed and developing states. The organization is run by an executive committee composed of seven developed and seven developing states. Every effort is made to make decisions by consensus. If that is not possible, decisions are made by "double majority voting," in which two-thirds of the members must vote in favor of a proposal, including majorities of both the developing and developed state representatives (DeSombre and Kauffman 1996, 99). The fact that the organization was created independently was a victory for developing states as well; developed countries preferred that the funding be handled through the World Bank, an organization that developing countries regarded with some suspicion.

What is also interesting is the extent of cohesion among developing states in this negotiating process. China, India, and some other large and rapidly developing states were the most important ones for those already involved in the Montreal Protocol to entice into the agreement. If these major developing states had been willing to accept a side deal for their participation, rather than the creation of a universally applicable institution to address developing country concerns, they might have been able to negotiate individually appealing compensation for their participation. The fact that they did not do so might be attributable to an effort at developing country solidarity, since at least part of the point of creating a funding mechanism was to codify the broader principle that developing states should not be asked to restrict their development activities in a way that already-industrialized states had not had to, without someone else bearing the cost. In any case, the precedent set by the multilateral fund has been critical: nearly every global environmental agreement negotiated since the Montreal Protocol has included funding for developing countries. Tying the requirement by developing states to meet obligations to the delivery by developed states of promised financial assistance has taken that precedent further in some agreements. For example, the United Nations Framework Convention on Climate Change (1992, article 4 [7]) specifies that "the extent to which developing country

Parties will effectively implement their commitments . . . will depend on the effective implementation by developed country Parties of their commitments . . . related to financial resources."

Time

One other unusual asymmetry in this case (but one that is likely to become more prevalent within emerging environmental issues) relates to time. The same stability of ozone-depleting substances that enables them to persist until they reach the stratosphere also grants them a long life once there: the most common CFCs survive for a century or more, and some can last for nearly two millennia (U.S. Environmental Protection Agency n.d.). This longevity means that emissions from previous generations affect the resource for those not even born at the time that the substances were created. These intergenerational problems are likely to cause increasing difficulties for addressing global environmental problems; they increase uncertainty (since it may take a long time between cause and environmental effect) and allow for the possibility that those actors that undertake the costs of addressing a problem may not themselves benefit from the environmental improvement. Under the best of circumstances, politicians have short time horizons and prioritize certain current benefits over uncertain ones that might accrue in the distant future (Gardiner 2004). This preference makes the efforts to address ozone depletion all the more impressive, since politicians agreed to impose costs on industry within their states despite uncertainty about the magnitude of the potential future problem, and even though they may no longer be in office when the environmental improvement is witnessed.

It also means that efforts to address the environmental problem now will not see immediate results. The substantial and dramatic reduction in emissions of ozone-depleting substances worldwide has not yet resulted in a recovery of the ozone layer, in large part because of the ozone-depleting substances emitted in the twentieth century that remain in the atmosphere. In addition to pushing effects on those who did not have a hand in the causes, this intergenerational inequity can make any resolution of the problem more complex.

Uncertainty, Degradation, and Cooperation

The issue of ozone depletion has been replete with uncertainties, the first of which was whether the problem even existed. Because ozone depletion was hypothesized before it was observed in the environment, there was initially legitimate uncertainty about the extent of the problem, and that uncertainty could be used for political ends by those who would be affected by potential regulation. On the issue of the degree of the problem, it is important to note that an unqualified understanding that ozone depletion was occurring and was attributable to human causes was not necessary for international cooperation to take place.

On the one hand, public opinion in favor of some form of regulation did seem to respond to the increasing scientific consensus that the problem was real and could have serious consequences, and it may be no accident that the strongest impetus for regulation came from within the United States, where the scientists who discovered the problem resided. On the other hand, the search for environmental evidence that the problem was real was not central to the willingness of states to negotiate international agreements. The seasonal Antarctic ozone thinning that has come to be known as the ozone hole was discovered in 1985, but at that point human responsibility for that thinning was not yet clear (Kerr 1986). It was not until 1988, shortly after the negotiation of the Montreal Protocol, that the Ozone Trends Panel attributed this phenomenon unquestionably to anthropogenic causes (Watson et al. 1988). In considering the question of whether a certain degree of environmental degradation is needed for, or may eventually become counterproductive to, international cooperation, the uncertainty of this case likely plays a role. The potential harm was great, but the actual evidence of it was small at the time that the initial cooperation took place.

It is nevertheless the case that cooperation both broadened and deepened as new knowledge emerged about the extent of the environmental problem. The initial abatement measures contained in the Montreal Protocol were modest compared to what would ultimately be undertaken: a 50 percent cut (by developed countries) relative to 1986 in CFC use by 1998 and a freeze on halons by 1992 (Montreal Protocol 1987 [unamended], article 2). As the potential seriousness of the problem became clearer, the parties agreed in amendments and adjustments to the protocol to increasingly strict and rapid controls on these substances, eventually resulting in the complete phase out of CFCs by developed countries as of 1996 and a complete phase out of halons by 1994. Restrictions on new substances were also added as their effect on the ozone layer was understood. Controls on carbon tetrachloride and methyl chloroform were added in 1990, and controls on hydrochlorofluorocarbons and methyl bromide were put in place in 1992 (see Parson

2003, 240–241). These changes did not happen without contention (most notably the continuing U.S. opposition to more controls on methyl bromide), but the increasing scientific understanding of the causes and consequences of ozone depletion made international cooperation to prevent it much less problematic than was the case in either 1985 or 1987, when the first international instruments were negotiated. The increasing awareness of greater environmental degradation, especially in the context of an already-functioning international agreement, led to deeper cooperation.

Conclusion

The international response to ozone depletion represents an optimistic case for international cooperation to address resource degradation. States cooperated internationally to address a problem of a degraded and shared resource. They were able to begin the process of cooperation before the extent of the problem was fully understood. Their cooperation deepened as it became clear that the problem was worse than originally feared. In addition, the process of cooperation included important support for developing countries that took on international regulation of ozone-depleting substances.

This case also demonstrates some broader lessons that fit in with the overall themes of this volume. An asymmetry of concern and responsibility (along with shrewd negotiating) rather than a general sense of global justice is what led to the funding mechanism to support developing countries as well as the decision-making process that met many of their concerns. But it is also the case that the location of greatest concern may not match with that of greatest harm from the environmental problem. In an issue where actions taken by one state affect the quality of a resource used by others, those with the least concern gain significant bargaining power. States that have—or can credibly claim to have—more pressing priorities than the environmental issue in question may gain the negotiating advantages, regardless of the environmental reality.

This issue, as should be apparent by now, has important differences with other issues of scarcity or environmental degradation; its global and diffuse nature makes it different from the type of resource use issues that underpin most of the environmental security literature. The issue shares characteristics of other common pool resource issues, however, in that participation by all that contribute to the degradation is necessary in order to address the problem. Moreover, the successful resolution of the ozone problem offers hope that international regulation can indeed, under some circumstances, address problems of resource degradation.

References

Barkin, J. Samuel, and George E. Shambaugh. 1999. "Hypotheses on the international politics of common pool resources." In *Anarchy and the Environment: The International Relations of Common Pool Resources*, ed. J. Samuel Barkin and George E. Shambaugh, 1–25. Albany: State University of New York Press.

Benedick, Richard E. 1991. *Ozone Diplomacy: New Directions in Safeguarding the Planet*. Cambridge, MA: Harvard University Press.

British Antarctic Survey. 2005. *British Antarctic Survey Bulletin* (December 19). Available at <http://www.theozonehole.com/ozonehole2005.htm> (accessed May 28, 2007).

Chasek, Pamela S. 2001. *Earth Negotiations: Analyzing Thirty Years of Environmental Diplomacy*. Tokyo: United Nations University Press.

Clapp, Jennifer. 1997. "The illegal CFC trade: An unexpected wrinkle in the ozone protection regime." *International Environmental Affairs* 9 (4): 259–273.

Clark, William C., and Nancy M. Dickson. 2001. "Civic science: America's encounter with global environmental risks." In *Learning to Manage Global Environmental Risks: A Comparative History of Social Responses to Climate Change, Ozone Depletion, and Acid Rain*, Vol. 1, ed. William C. Clark, Jill Jäger, Josee van Eijndhoven, and Nancy M. Dickson, 259–294. Cambridge, MA: MIT Press.

DeSombre, Elizabeth R. 2001. "The experience of the Montreal Protocol: Particularly remarkable and remarkably particular." *UCLA Journal of Environmental Law and Policy* 19 (2): 49–81.

DeSombre, Elizabeth R. 2006. *Global Environmental Institutions*. London: Routledge.

DeSombre, Elizabeth R., and Joanne Kauffman. 1996. "The Montreal Protocol Multilateral Fund: Partial success story." In *Institutions for Environmental Aid*, ed. Robert O. Keohane and Marc A. Levy, 89–126. Cambridge, MA: MIT Press.

Downie, David. 1999. "Ozone depletion and common pool resources." *Anarchy and the Environment: The International Relations of Common Pool Resources*, ed. J. Samuel Barkin and George E. Shambaugh, 97–121. Albany: State University of New York Press.

Friends of the Earth. 1990. *Funding Change: Developing Countries and the Montreal Protocol*. Amsterdam: Friends of the Earth.

Gardiner, Stephen M. 2004. "The global warming tragedy and the dangerous illusion of the Kyoto Protocol." *Ethics and International Affairs* 18 (1): 23–39.

Kerr, Richard A. 1986. "Antarctic ozone hole is still deepening." *Science* 232 (June): 1602.

Molina, Mario J., and Sherwood F. Rowland. 1974. "Stratospheric sink for chlorofluoromethanes: Chlorine atom-catalyzed destruction of ozone." *Nature* 249: 810–812.

Montreal Protocol on Substances That Deplete the Ozone Layer. 1987. Available at <http://ozone.unep.org/Publications/MP_Handbook/Section_1.1_The_Montreal_Protocol/> (accessed May 28, 2007).

NASA Advanced Supercomputing. 2001. "The Antarctic ozone hole." June 26. Available at <http://www.nas.nasa.gov/About/Education/Ozone/antarctic.html> (accessed May 28, 2007).

Ozone Secretariat. n.d. "2002 findings of the assessment panel." Available at <http://hq.unep.org/ozone/Public_Information/4Av_PublicInfo_Facts_assessment.asp> (accessed May 28, 2007).

Parson, Edward A. 2003. *Protecting the Ozone Layer: Science and Strategy.* Oxford: Oxford University Press.

Sprinz, Detlef, and Tapani Vaahtoranta. 1994. "The interest-based explanation of international environmental policy." *International Organization* 41 (1): 77–105.

United Nations Framework Convention on Climate Change. 1992. Available at <http://unfccc.int/essential_background/convention/items/2627.php> (accessed May 28, 2007).

U.S. Environmental Protection Agency. n.d. "Class I ozone-depleting substances." Available at <http://www.epa.gov/docs/ozone/ods.html> (accessed May 28, 2007).

Vienna Convention for the Protection of the Ozone Layer. 1985. Available at <http://ozone.unep.org/Publications/VC_Handbook/Section_1_The_Vienna_Convention/index.shtml> (accessed May 28, 2007).

Watson, Robert T. and Ozone Trends Panel, Michael J. Prather and Ad Hoc Theory Panel, and Michael J. Kurylo and NASA Panel for Data Evaluation. 1988. "Present state of knowledge of the upper atmosphere 1988: An assessment report." *NASA Reference Publication* 1208. Washington, DC: U.S. Government Printing Office.

4

Biodiversity Protection in International Negotiations: Cooperation and Conflict

G. Kristin Rosendal

This chapter focuses on the issues of loss and scarcity related to biological diversity and the value of biodiversity. Biological diversity is a broad concept that includes the variability among all species and ecosystems as well as the diversity within species—the genetic diversity.[1] The loss of biodiversity has ramifications for a broad spectrum of ecosystem services, as the issue comprises several levels and sectors (Millennium Ecosystems Assessment 2005). These include conservation and management at the ecosystem level, various forms of protection at the species level (relating to trade or migration), and the controversial issue of access and benefit sharing (ABS) versus intellectual property rights at the level of genetic resources.

The chapter commences with a brief historical overview of the topics at these various levels, demonstrating how degradation culminated in international cooperation. Agreements such as the Ramsar Convention on Wetlands and the Convention on International Trade in Endangered Species of Flora and Fauna (CITES) are cited. These initiatives can be otherwise termed piecemeal attempts at cooperation.

Above all, this chapter discusses the Convention on Biological Diversity (CBD), along with its comprehensive approach to biodiversity protection at all levels and sectors. Beyond biodiversity loss and value, an examination of the other factors that hampered and facilitated the CBD negotiations and outcome is also provided. Multiple levels of power asymmetries and controversies as well as evolving political norms and principles, for example, were crucial to understanding the final negotiated outcome of the CBD.

The final part of the chapter looks beyond the cooperative solution, and dwells on the ABS issue in the implementation phase. While the ABS issue can be seen as a great success for cooperation at the normative level (the CBD likewise contains important elements on how to deal with the

ABS conflict in principle), implementation has been difficult. Asymmetries remain a stumbling block, power is often a critical component in decision making, and cooperation is far from accomplished in day-to-day policy.

The Multifaceted Issue of Biodiversity: Scarcity and Piecemeal Cooperation

When discussing cooperation on biological diversity, it must be pointed out that the concept itself is quite recent, as it surfaced in the 1980s (Wilson 1988). Still, early efforts at biological diversity protection and awareness of species loss are far from novel. Human-induced species loss was first recognized with the extinction of the big, flightless bird, the dodo (*Raphus cucullatus*), in about 1650 (Quammen 1996). With regard to conservation efforts, the United States has been a pioneer by establishing the world's first national park, Yellowstone, in 1872.

Efforts at formal international cooperation have also preceded the CBD as a response to biodiversity loss—albeit at a less comprehensive level. These are less easily identified, however, as some of the earliest conservation activities came out of cooperation on natural resources management, such as the International Whaling Commission (established in 1946). The North Pacific Fur Seal Treaty of 1911, responding to the hunting and subsequent loss of fur seals, is seen as one of the most successful international environmental treaties. It provided a transparent system of incentives and enforcement that improved management along with the payoffs of the parties (Barrett 2003). A central characteristic of these early management regimes is that the users were cooperating on improving the management of a resource in order to create integrative results—that is, win-win situations, where all the participants might obtain a larger share as a result of cooperation (Underdal 1980).

During the 1970s a set of international conventions were likewise negotiated. These initiatives focused on the protection of specific endangered species or habitats. Those with a global scope include most significantly the Ramsar Convention on Wetlands (1971), the Bonn Convention on Conservation of Migratory Species (CMS 1979), the Convention concerning the Protection of the World Cultural and Natural Heritage (WHC 1972), the Washington Convention on International Trade in Endangered Species of Wild Flora and Fauna (CITES 1973), and the Food and Agriculture Organization's (FAO 1983) International Undertaking on Plant Genetic Resources.[2] Many of these were partly motivated by the 1972 UN Conference on the Human Environment in Stockholm (Lanchbery 2006). In addition, there are a sizable number of regional agreements, such as the Convention on the Conservation of European Wildlife and Natural Habitats (1979) and the African Convention on the Conservation of Nature and Natural Resources (1968). Considering some of these initiatives in more detail is instructive.

- CITES seeks to protect endangered wild species by regulating trade. Because the trade in wild animals and plants crosses borders between countries, efforts at regulation required international cooperation to safeguard certain species from overexploitation. Today, CITES accords varying degrees of protection to more than thirty thousand species of animals and plants, whether they are traded as live specimens or manufactured goods, such as fur clothing, elephant tusks, and herbal medicine. The most controversial issues concern those where relatively poor owners of a resource are being asked to refrain from utilizing a scarce but valuable resource through trade, such as the case of elephant tusks in southern Africa.
- The CMS aims to conserve endangered terrestrial, marine, and avian migratory species throughout their range. The CMS is a response to scarcity and loss in migrating species, which could not be resolved within the policy borders of single countries. It is global in scope, acting as a framework convention, whereby several regional agreements have subsequently been concluded (i.e., for bats, cetaceans, albatrosses, and seals). The UN Environment Programme (UNEP) is directly involved (in negotiating and implementing) some of the CMS-related agreements, such as the Agreement on the Conservation of African-Eurasian Migratory Waterbirds (AEWA 1995) and the Agreement on the Conservation of Populations of European Bats (EUROBATS 1991). The CMS Agreement on the Conservation of Gorillas and Their Habitats (2007) includes the development of financing and fund-raising mechanisms to permit the implementation of this transborder action plan.
- The Ramsar Convention on Wetlands provides the framework for national action and international cooperation for the conservation and wise use of wetlands and their resources. It is the only global environmental treaty that deals with a particular ecosystem, and includes 1,700 wetlands and covers more than 153 million hectares. Ramsar is a cooperative response to the loss of wetlands with the acknowledgment that this causes serious and sometimes irreparable environmental damage to the provision of ecosystem services.

Several other conventions and programs could be mentioned, such as regimes dealing with fisheries management and the marine environment. A full listing of all types of cooperative arrangements in the field relating to biodiversity would greatly exceed the scope of this chapter, though. Suffice it to say, UNEP is central in the establishment and organization of the bulk of international biodiversity activities, providing secretariat functions for CITES, the CMS, and as will be discussed later, the CBD. UNEP also manages a number of conservation programs and projects. Since there is not a world government to enforce compliance with environmental agreements, UNEP is faced with a formidable task (Andresen and Rosendal 2007). This makes the incentive structure within the treaties all the more important.

In common for many of these cooperative responses is that the owners of the scarce resources are to some extent compensated for restricting their use of the resources. Since most of the terrestrial species diversity is located in developing countries in the South and the major concern for biodiversity loss is voiced in the North, the North-South conflict runs through all of these conventions.[3] Another challenge to the more traditional conservation treaties, such as CITES, is the increasing emphasis on the social dimension of conservation, along with issues of legitimacy and equity (Hutton and Dickson 2000).

During the 1980s it was becoming apparent that despite their utility, the piecemeal conservation treaties were not enough to stem the loss of biodiversity worldwide. This led to the next generation of cooperative responses to the loss of biodiversity—most notably the CBD

Barriers and a Breakthrough in Cooperation on the CBD

Needing the CBD: Scarcity and Value

It was not until the late 1980s that the global issue of biodiversity loss hit international agendas (Wilson 1988). There are varying estimates as to the total number of species, with additional uncertainties regarding how the loss of single species affects ecosystems and ecosystem services. There might be between 7 and 30 million species, of which some 1.9 million have been described scientifically. Much more scientific certainty exists regarding the seriousness of biodiversity loss, which is estimated at about a hundred times the natural background rate—that is, without human intervention (Heywood 1995, 232). Science is also in agreement that the main human-induced drivers of this loss are the fragmentation and deterioration of habitats, alien invasive species, pollution, and climate change. Coinciding with the start of the CBD negotiations, a World Resources Institute (WRI) report (Reid and Miller 1989) warned that a quarter of the world's species may be gone by 2050. The authors provided evidence that the biodiversity crisis was not restricted to tropical forests but threatens biological resources in temperate zones and marine ecosystems as well. Echoing the message of the World Commission on Environment and Development (WCED 1987), it was also acknowledged that the long-term actions called for would not occur in the developing world unless the industrial countries supply their fair share of the financing, technology, and knowledge needed to implement them. The WRI report was followed by several compilations of studies by the WRI, the World Conservation Union, and UNEP (McNeely et al. 1990; WRI et al. 1992), which along with the biennial *World Resources* and *Environmental Data* and annual *UNEP State of the Environment* reports, provided authoritative and disturbing overviews of the state of the planet's biodiversity.[4]

Biodiversity as a whole provides a great range of ecosystem services, such as local/regional water and climate regulation, soil protection, pest control, and crop pollination. Biodiversity likewise supplies food, fodder, and materials for building and firewood (Martens, Rotmans, and de Groot 2003).[5] There is also a great range of noneconomic values attached to biodiversity, such as cultural, spiritual, recreational, and other intrinsic values (Millennium Ecosystem Assessment 2005; Myers 1996). Since 1900, about 50 percent of the world's wetlands have disappeared (Moser, Prentice, and Frazier 1996) and some 30 percent of coral reefs have been seriously damaged (Wilkinson 2004). Over the last two decades, 35 percent of mangroves have disappeared (Millennium Ecosystem Assessment 2005).

The increasing economic value of genetic resources is also significant. As new biotechnologies were emerging (new in the sense that the technology was moving beyond traditional breeding, brewing, and baking), estimates pointed to the growing value of genetic resources to biotechnology. Wild relatives of domestic crops, for example, provide genetic variability that can be crucial for overcoming disease outbreaks as well as adapting to climate change or other threats (Kloppenburg and Kleinman 1987; Kloppenburg 2004). In addition, access to genetic resources is necessary for further development in all plant and animal breeding. At the same time, there is a rapid loss of genetic diversity in domesticated plants, with potential risks for food security (FAO 1998). Biodiversity conservation is similarly valuable because species may contain compounds that can generate pharmaceuticals or other products. It has been estimated that U.S. biotechnology produces an annual revenue of US$13 billion (Chambers 2002), and the estimated value of products derived from genetic resources worldwide is between US$500 and US$800 billion (ten Kate and Laird 1999; Bishop et al. 2008). Other scholars argue that the economic value of biodiversity is rather low because it remains plentiful relative to the demand for it (Simpson, Sedjo, and Reid 1996).

"Bioprospecting," the screening of biodiversity in search of commercially valuable genetic and biochemical resources, is also useful to biotechnology—often taking the form of plants with medicinal traits, chemicals such as colors and enzymes, and construction materials used in the industry sector (WRI 1993).[6]

As suggested above, a broad political and scientific consensus confirmed the need to halt biodiversity loss. The role of scientific uncertainty with regard to the biodiversity case is much more concerned with the number of species and not the extent of rapid loss.[7] The loss of biodiversity was therefore a chief motivator for the international action that culminated in the CBD. Other barriers of a more political nature, however, impeded cooperation efforts. Nonetheless, these were generally overcome in the CBD negotiations.

Main Barriers: Asymmetries in the CBD Negotiations

The novel trait of the CBD is its comprehensive scope, including all species and ecosystems worldwide, as well as the genetic diversity within species. The CBD is even more novel in that it includes equity aspects along with conservation initiatives—the equitable sharing of benefits from the utilization of genetic resources. The formula negotiated would seem quite a feat once the asymmetries inherent in the biodiversity case are realized. In fact, considering the innate distributive aspects of the biodiversity case, ultimately a result of the property rights dispute (and the dynamics between the users and owners of the resource), particularly illuminates the convention's achievement in the face of such contextual obstacles.[8]

An estimated 95 percent of the world's terrestrial species diversity is found within the jurisdiction of sovereign states, and the major bulk of species diversity predominates in the South (Heywood 1995). This would seem to indicate that the main issue of the CBD negotiations would not involve a commons resource but also one with well-defined property rights. Regardless, the negotiating parties had different views regarding this classification. Historically, plants and micro-organisms that originate in the South have been collected free of charge and then stored in international gene banks in the North (Kloppenburg 2004; Barton and Siebeck 1992). Biotechnology companies, largely from the North, also access genetic resources as well as traditional knowledge about valuable medicinal traits through bioprospecting in the South.

Traditionally, genetic resources have been associated with the *common heritage of humankind* concept, and they were treated this way within the FAO negotiations on a nonlegally binding International Undertaking on plant genetic resources in the 1980s.[9]

Broadly speaking, then, the North and South started out with vastly different environmental agendas for negotiation of the biodiversity issue. Some of the more powerful countries of the North set out to obtain an international commitment to conserve wild species and habitats, and to leave their free access to the biological resources of the South out of the CBD text (Koester 1997; Rosendal 2000). The objective of conservation was explicitly advocated, and was spurred on by rain forest destruction and other threats to biodiversity worldwide. The objective of free access to genetic resources may thus be seen as a "hidden agenda," and reflected the increased economic and biotechnological interest in biological material.

The South, in contrast, wanted to be compensated by the developed world for the costs involved in conservation activities—the costs of refraining from utilizing their own resources. Second, many developing countries wanted to include domesticated materials on the CBD agenda, as this highlighted the economic benefits accruing from the access to and use of genetic materials. Third, the South wanted access to the same technologies that enable industrialized countries to profit from the exploitation of biological resources. The South's main objection to the negotiations was that the (primary) users of the resource were asking the (primary) owners to change their behavior—to conduct resource conservation and submit to stricter regulations.

In effect, the evolving property rights regime immediately preceding the CBD negotiations gave rise to asymmetries between actor interests along two dimensions and led to cumulative asymmetry. The international distribution of biodiversity itself is uneven, as are the capabilities to utilize biological resources economically and technologically, and this leads to a double asymmetry:

• Economic benefits for users: The countries with relatively low levels of biodiversity frequently have a higher capacity to benefit from its use, due to high technology levels, and vice versa.

• Economic costs for owners: Conservation costs will place a heavy burden on the owner countries, and will have adverse effects on sectors such as agriculture and forestry, on which these countries are heavily dependent.

Developing countries therefore protested and argued that their genetic capital was still considered a common heritage of humankind, which is freely accessed by all, while they themselves have to pay dearly for access to improved breeding materials and medicines (Rosendal 1991, 2000; Raustiala and Victor 2004). The classical example is the plant rosy periwinkle found in Madagascar. The plant is the source of a widely used medicine for leukemia, and the multinational corporation Eli Lily patented the active ingredients. The drug reportedly generated an annual US$200 million, none of which was returned to the country of origin (Wilson 1992, 283; ten Kate and Laird 1999). Other examples that have generated similar conflicts are the Indian neem tree and the hoodia plant that is traditionally used by the San people in the south of Africa. The broader picture shows that traditional knowledge increases the success ratio of bioprospecting by 400 percent (Gehl Sampath 2005). In effect, both groups of actors were drawn to the CBD negotiation table with great concerns, but with quite different agendas in mind.

Evolving property rights regimes were further complicating this situation, however. Most important, the World Trade Organization (WTO) negotiation process—the Uruguay Round of the General Agreement on Tariffs and Trade from 1988 to 1994—coincided with that of the CBD, and was partly aimed at strengthening and harmonizing patenting in all technologies, including biotechnology. Patenting and intellectual property rights introduced an emphasis on private rights over certain types of genetic resources, and hence caused a breach with the traditional approach of common heritage.

The development of modern biotechnology coincided with the increased privatization of agricultural and pharmaceutical research in the 1970s. This brought about a need for—and the application of—patents. Biotechnology made it possible to fulfill the legal patent criteria for inventions involving biological material (Crespi 1988).[10] At the same time, it has proved difficult to provide similar legal protection of the traditional knowledge about these resources. Patenting is also a costly affair, largely dominated by multinational corporations (Gleckman 1995). In the late 1980s, developing countries held only about 1 percent of all patents in biotechnology, and by 2005 that figure had only increased to 4 percent (WCED 1987; UNDP 2005, 135).

The result of these new trends of patenting in biotechnology has been a gradual change in the overall approach to property rights to genetic resources—from the nonexclusiveness (based on the common heritage principle) of all genetic materials to exclusive rights (i.e., intellectual property rights like patents) for parts of genetic materials (Koester 1997; Rosendal 2000).

Initially, the UNEP negotiation agenda for the CBD was dominated by a focus on classical wildlife conservation—protecting biological hot spots such as tropical rain forests and other places of high biological diversity. The developed countries feared that the link to domestication, and hence systematic or nonsystematic breeding, would inevitably lead back to the controversial issue of property rights and compensation, as the preceding FAO negotiations had already shown. The debate and controversy were fueled even more by the parallel General Agreement on Tariffs and Trade negotiations on the Trade-Related Aspects of Intellectual Property Rights (TRIPS 1994) agreement, which also included biotechnology—and hence, property rights to genetic resources (Rosendal 1991, 2000; Porter 1993). In response to the ongoing WTO/TRIPS negotiations on intellectual property rights, the developing countries started advocating the principle of national sovereignty over biological resources (Rosendal 2006a).

In was only in the last negotiation round that the CBD's threefold objective—conservation, sustainable use, and equitable sharing—was accepted, along with the principle of national sovereignty. These combined efforts accommodated the call for a common responsibility for biodiversity conservation as well as compensation to the countries providing biological resources. In response to the common responsibility and compensation question, the parties also agreed on the Global Environmental Facility (GEF) as the interim financial mechanism of the CBD.

The Negotiation Result

The CBD was signed at the UN Conference on Environment and Development (UNCED) in Rio de Janeiro in 1992, and entered into force the following year. One hundred and ninety-three states have ratified and/or are parties to the CBD—among the largest number of parties to any global treaty (CBD 1992).[11]

The CBD sets out obligations and objectives for nations to combat the destruction of plant and animal species and ecosystems. For this purpose, the contracting parties are instructed to develop national strategies, plans, and programs for conservation and sustainable use. The parties are supposed to integrate conservation and the sustainable use of biological diversity into relevant sector plans and policies, and develop systems of protected areas (CBD 1992, article 6). They are also obligated to identify components of biological diversity that are important for its conservation and sustainable use, monitor the components through sampling and other techniques, and identify activities that have or are likely to have significant adverse impacts on conservation and the sustainable use of biodiversity. Consequently, the parties are likewise instructed to introduce environmental impact assessments with a view to avoiding or minimizing these effects, and establish a system of protected areas (in situ conservation).

The international community is given responsibility for conserving biodiversity in developing countries (CBD 1992, article 8[m]) through new and additional financial resources (CBD 1992, preamble and article 20.2). The CBD stipulates that developing countries must implement their obligations on conservation and sustainable use to the extent that developed country parties meet their commitments related to financial resources and technology transfer (CBD 1992, article 20.4).

The CBD's objective with regard to conservation has been to halt the loss of biodiversity by 2010.[12] With a view toward the use and equitable sharing of benefits, the CBD seeks to counterbalance the expanding patent regime by exchanging the common heritage principle with that of national sovereignty over genetic resources (Koester 1997; Rosendal 2000).

The CBD was largely formulated as a compromise between access to technology and access to input factors—genetic resources. Hence, in article 16.5 the parties are asked to ensure that intellectual property rights do not run counter to the CBD objectives. Through article 15.7, the parties have accepted the aim of "sharing in a fair and equitable way the results of research and development and the benefits arising from the commercial utilization of genetic resources with the Party providing such resources." Article 15.2 encourages the parties to "create conditions to facilitate access to genetic resources for environmentally sound use." Without access, it is presumed there will be fewer benefits to share. Without benefit sharing, it is suggested that there will be fewer resources conserved for future use. Prior informed consent (article 15.4) and mutually agreed-on terms (article 15.5) are therefore meant to guide access to genetic resources. Since signing the CBD in 1992, the parties have continued negotiating access and equity, first resulting in the voluntary Bonn Guidelines in 2002 on ABS.[13]

The Bonn Guidelines encourage prior informed consent and mutually agreed terms by making concrete suggestions for how these principles may be included in bilateral bioprospecting agreements. They suggest elements for Material Transfer Agreements—that is, a document recording the prior informed consent and all terms under which genetic resources are acquired, and emphasize the involvement of stakeholders. They encourage disclosure of the country of origin in patent applications, propose a certification system for trade in genetic resources, and suggest the establishment of an international ombudsperson for monitoring infringements on bioprospecting deals.

There are certainly flaws and limitations in the CBD. For instance, the Bonn Guidelines are not legally binding. This is still a contested issue, and CBD members are now seeking to transform the guidelines into legally binding language on ABS. Moreover, the CBD text is largely framed in symbolic terms, as it proved difficult to set any technical standards pertaining to any of the three objectives (conservation, sustainable use, and equitable sharing). A common trait of many goals in relation to nature management is that there are few explicit standards by which to measure success or failure in terms of implementation.[14]

In all, at the time that the CBD was concluded it was seen as a great victory for the developing countries' negotiation agenda despite strong opposition from the developed world. Finally, while increasing scarcity and loss in biodiversity indeed prompted the CBD negotiations, the inherent asymmetries could easily have torpedoed this initiative. Evolving political norms and the use of side payments were likewise important to understanding the convention's negotiation.

Beyond Scarcity: Explaining the CBD through Side Payments and Evolving Norms

The rapid loss and scarcity of biodiversity was necessary but hardly sufficient to induce cooperation on an issue that was plagued by asymmetries in the negotiation phase. The use of compensation and side payments combined with evolving norms and a changed political climate played a role as well in prompting agreement on conservation and an equitable sharing of benefits.

Recalling the basic power and interest structure that characterize the biodiversity case may suggest that the South had a relatively strong bargaining position (despite its relatively weaker position in terms of absolute power) in the international negotiations leading up to the CBD text.[15]

By looking at the two dimensions of conservation and equitable sharing, and the interests tied to them, we may discern a tendency toward a more symmetrical relationship between the North and South in the biodiversity issue area. Particularly, the linking of these dimensions signified an evolution from an asymmetrical interest structure to a semblance of symmetry between the actors' preferences. As the North and South came out as pushers on the respective dimensions, a certain complementarity in interests seemed possible through side payments. It could be argued that the North's interest in securing access to genetic material has given the South an increased issue-specific power. In fact, the majority of developed countries eventually realized that disregarding the equity issues would mean that the developing countries, especially those such as Brazil and Malaysia, would not join the CBD (Schei 1997). As the bulk of biodiversity is located in the tropics, negotiations could hardly proceed without the South. In response to the property rights question, it was agreed that the issues of access and control over biological resources could not be treated separately from the ongoing WTO/TRIPS negotiations.

Evolving norms and a changed political climate also played a role in the CBD's successful negotiation.

First, timing was a factor. The CBD's negotiation largely coincided with the UN Conference on Environment and Development (UNCED) in Rio in 1992. Global public attention and public pressure pushed the top politicians to act. Even though a recurring criticism from the South was that UNCED was concerned with the environmental agenda of the North, developmental issues were nevertheless more overt in this forum than in the preceding UN Conference on the Human Environment in 1972. The undercurrent of normative persuasion became linked to the recognition that these resources did indeed belong to the South. In this light, the effect of the institution, in terms of its official goal of framing the issue, should not be overlooked (Rosendal 2000). While such sentiments would probably have less impact in forums on economy and trade, the goals associated with the biodiversity negotiations in UNEP may have been more receptive to accept the legitimacy of granting the countries of the South their "rights."

Second, when the CBD was adopted and signed, environmental issues loomed larger on most political agendas than previously recorded. During the two decades that led up to the Rio conference, the majority of states, both developed and developing, had acquired some form of environmental agencies. Moreover, there was an unprecedented growth of nongovernmental environmental organizations, both at the domestic and international level, and these groups were increasingly granted observer status in international environmental negotiations. States were becoming increasingly aware of their environmental reputation, and only a few were indifferent as to whether they appeared as laggards in international environmental cooperation (Rosendal 2000).

Third, the inclusion of all relevant parties is an important point, especially as it pertains to the overall political climate of negotiations. All states are potentially relevant parties to the collaboration on biodiversity conservation, but as the number increases, the scope for successful cooperation and implementation is presumably reduced as the risk of the other parties' free riding increases. In the biodiversity area, one solution might have been to concentrate on the participants representing the largest donors (as conservation will be costly) and the so-called megadiversity countries (with high levels of biodiversity endemism). The political danger in this approach was that this would have put the main responsibility for conservation on the megadiversity countries of the South. Hence, it was deemed politically unfeasible to limit the CBD to issues of wildlife conservation. The much more disputed aspects related to the utilization of biological resources were added to the agenda—thus, compensation to the South became a topic (Schei 1997). This all-encompassing approach also increased the scope for issue linkage, thereby enhancing the legitimacy of the agreement with all the parties—developed and developing alike. In sum, the number of participants seems to have broadened the scope for achieving more integrated solutions.

Finally, the available scientific knowledge on biodiversity loss bolstered the political climate in favor of cooperation. Critical scientific findings regarding *Waldsterben* (or forest death) and loss of biodiversity through rain forest destruction—all in the context of transborder environmental problems—led to an increasing public awareness that action was necessary along with the realization that concerted international action was crucial too. In addition, stories such as the rosy periwinkle, neem, and hoodia continued to anger the developing countries in the biodiversity negotiations, and the mood grew steadily in favor of securing an equitable share for poor countries. It was also recognized that the developing countries could not be expected to pay the full costs of conserving biodiversity. In this sense, it was seen as a victory for the developing world that an agreement on this convention was achieved at all. The convention did meet many of the South's demands—at least in writing.

The prospects for implementation of the CBD objectives are paramount, as a result, and are considered in the following section. Such a process does not take place in a vacuum. This is why scholars studying the implementation of multilateral environmental agreements increasingly take regime interaction into account (Oberthür and Gehring 2006; Rosendal 2001a, 2006a; Young 1996). The idea of studying interaction is based on the notion that the effective implementation of one convention may affect the implementation of another.[16] A better understanding of the synergies and conflict potential relating to interaction is thus crucial for achieving improved implementation. The influence of the FAO and TRIPS during the negotiation phase of the CBD has already been mentioned. Yet these effects are also relevant in the implementation phase. The FAO and TRIPS—which compete with the CBD objectives for conserving and sharing benefits from genetic resources—are presented below in the context of the implementation analysis.

Cooperation in the Implementation Phase: Sisyphus at Work

Questions explored in the following section is whether the much-discussed ABS component of the CBD is more than a symbolic victory for the developing world, and whether it is actually being implemented. Another issue investigated here is whether the parties have progressed with the goal of halting and reversing the loss of biodiversity by 2010.

The State of Implementing the CBD Objectives

In assessing the state of biodiversity conservation several findings are relevant. The Millennium Ecosystem Assessment (2005) report, for example, concluded that the world, particularly the poor countries, are not reaching the objectives of the conservation and sustainable use of biodiversity. While rates of habitat loss are decreasing in temperate areas, they are projected to continue to increase in tropical areas (Millennium Ecosystem Assessment 2005, 14). On the one hand, the percentage of protected areas has doubled from 6 or 7 percent to 12 percent since 1990. On the other hand, the quality of protection is still contested (Millennium Ecosystem Assessment 2005, 11). The assessment concludes that biodiversity loss deprives people of ecosystem services worth about US$250 billion yearly. The only ecosystem services that are increasing globally are agriculture, temperate forestry, and aquaculture.

The most important multilateral financial institution in this regard is the GEF, as it is aimed at "[helping] developing countries fund projects and programs that protect the global environment."[17] So far, projects on the conservation and sustainable use of biodiversity have received GEF funding on the scale of US$2.6 billion, but as the Millennium Ecosystem Assessment showed, the GEF has not succeeded in stemming the rapid loss of biodiversity. Looking to cofinancing opportunities, the private sector is more interested in climate projects, as these provide a greater scope for identifying win-win aspects, at least on a short-term basis.[18]

It seems safe to conclude that without progress on the equity dimension to boost the legitimate will in developing countries, the goals for biodiversity conservation can hardly be successfully implemented. This brings us back to ABS and the fact that there has so far been little evidence of improved equity in bioprospecting deals. Moreover, as this chapter goes to press, the CBD parties have yet to agree on a legally binding ABS regime to replace the nonlegally binding Bonn Guidelines of 2002. The lack of movement on this dimension is certainly not the only reason why the CBD goal of stemming the loss of biodiversity is not being achieved. The amount of funding available for biodiversity conservation is, after all, only a fragment of what has been estimated from a great number of sources as necessary for halting and reversing the loss of biodiversity.[19] Yet the ABS issue points back to the asymmetries that hampered the negotiation phase in the first place, and hence is a rational place to start looking for ways out of the quagmire that the CBD is presently in.

Legitimizing Implementation: Access and Benefit Sharing

The CBD, TRIPS, and the FAO embody objectives that are internationally acclaimed and agreed on. Still, these objectives are often incompatible per the ABS issue (Rosendal 2006b).

Breeders and bioprospectors require the legal protection of genetic materials in order to ensure an economic return from their investments. This need is acknowledged and provided for in the TRIPS objectives. At the same time, farmers, breeders, and bioprospectors need access to improved breeding material—in order to carry out innovations and improvements for food and medicine (Correa 1999; Andersen 2008). This need is acknowledged and provided for in the objectives of the International Treaty on Plant Genetic Resources for Food and Agriculture (2001).[20] At the same time, the CBD promotes biodiversity conservation, and the CBD parties have agreed that the equitable sharing of benefits from the use of genetic resources is an important incentive for biodiversity conservation. In all, it is the expansion of the patent system into the area of biotechnology that has led these regimes to interact (Tolba 1998, 144).

This raises the question of how a balance can be achieved between access, equity, and legal protection in order to stimulate both innovation and conservation. Such a balance could, for example, imply that legislation in user countries should be compatible with legislation in countries providing genetic resources. Currently, the genetic resources and traditional knowledge about the resources of the South are still predominantly seen as a commons, even as intellectual property rights systems are being strengthened worldwide.

On the one hand, the TRIPS agreement leaves it to the WTO member states to decide on how to take environmental and socioeconomic considerations into account. Plants and animals may be exempted from patenting. On the other hand, the WTO Doha round is currently at a stalemate, and the trend is for developed countries to enter into bilateral free trade agreements with developing countries. These bilateral free trade agreements are often referred to as "TRIPS plus" agreements as they demand stronger and wider patent protection compared to TRIPS (Dutfield 2001). TRIPS plus agreements tend to limit the use of the compulsory licensing of medicine, and in several cases they do not allow for the exclusion of plants and animals from the patent laws of signatory countries. This threatens access to affordable medicines and valuable breeding material for poor countries (Choudry 2005).

Another important political forum is the World Intellectual Property Organization (WIPO). WIPO has 180 member states, which are currently negotiating the standardization of patent criteria in the members' domestic legislation. There seems to be little inclination among the developed countries to open up for additional patent criteria, such as the proposal on the disclosure of origin of genetic materials. Disclosure may increase the potential for equity in bioprospecting deals, partly because it can improve the search for *prior art* in order to avoid illegitimate acquisitions of traditional knowledge (Tvedt 2005; WIPO 2005).[21] Within the current system, it is legally unclear whether traditional knowledge (e.g., about the medical properties of a certain plant) constitutes prior art. Hence misappropriations occur (WIPO 2005). In effect, it is argued that patenting may be incompatible with the CBD policy objective of the equitable sharing of benefits (Hendrickx, Koester, and Prip 1993; Correa 1999).

The reaction from many developing countries has been to establish legal domestic regimes for ABS pertaining to genetic resources in order to secure an equitable share from their use. There are now fifty-eight countries and a number of regions that are in the process of establishing, or have already adopted, ABS measures—but these are exclusively developing countries.

As this chapter goes to press, the prospects for a legally binding and implementable multilateral system on ABS still seem to be distant, and developing countries in general have little negotiation power in bilateral forums.[22] First, only a few have the necessary administrative capacity to monitor bioprospecting deals. Genetic resources are not sold in bulk in the way that oil and minerals are; their value lies in qualitative traits. The most-cited example is the bioprospecting deal between Costa Rica and Merck and Co. Benefit sharing and technology transfer constitute core elements of this package deal as well as training and conservation—but the exact number and content of similar deals are unknown. Still, while benefit sharing clearly differs from one sector to another, the CBD standards for benefit sharing have become widely accepted (Laird 2000; Laird and Wynberg 2008, 117). Second, the South's negotiation power in bioprospecting deals is likely to remain low due to the fact that biological material does not necessarily follow political borders (Schroeder 2000). This makes it hard to establish who the beneficiaries of bioprospecting deals should be—that is, Which is the country of origin? and Who should benefit? (Fowler 2001). A third problem, and the central question, is tied to whether the gains to be made from bioprospecting deals will ever add up to the expectations, let alone meet the need for funding conservation. On a similar note, the activity of establishing ABS legislation in developing countries is criticized for being too cumbersome. It is claimed that this type of regulation will reduce access to important genetic materials for plant breeding and medicines, and also that access regulation could potentially endanger conservation as the scientific examination of tropical species is hindered (Grajal 1999). Thus, the question arises as to whether access to genetic resources is about to displace the goal of conservation.

Whether there may be (sufficient) benefits to be shared will partly depend on the level of interest in access to genetic resources. In a broad study of the varying needs and interests for wild genetic resources among biotechnology and seed sector companies, Sarah Laird and Rachel Wynberg (2005; 2008, 16) found that the pharmaceutical sector is experiencing a dwindling interest in and reliance on natural resources, due to the increased use of a synthetic, chemical approach and a reduced emphasis on infectious diseases. They argue that this may well change, however, as the fight against infectious diseases is not entirely over. Similarly, for industrial biotechnology, natural resources dependency is expected to resurface, as natural enzymes are biodegradable and hence more environmentally sound than synthetic chemicals. The seed and crop protection sector, in particular, still uses wild genetic resources (Laird and Wynberg 2005, 22), although there is uncertainty about the future value of these resources. In summary, across the board there is no conclusive evidence that the various sectors will not experience an increased need for access to natural, wild genetic resources. This could rekindle the bargaining power of developing countries, as they have sovereignty over the bulk of the resources.

On the other hand, there is growing interest in conducting bioprospecting in the North. The pharmaceutical sector increasingly turns to prospecting, say, unknown marine resources in temperate areas and is also bioprospecting closer to home (Haefner 2003). Here, the multinationals face less stringent ABS regulations, as no developed country has established strict ABS regulations. The trend could seem to bolster the contention that ABS legislation hampers access, and that the various parts of the biotechnology industry might choose to do business elsewhere and avoid the problem altogether. Yet Laird and Wynberg's findings (2005, 2008) indicate that it is the cumbersome process and lack of administrative capacity surrounding ABS legislation that creates mistrust in the industry sector, rather than the idea of equity itself. They maintain that the principle of the equitable sharing of benefits is increasingly accepted among larger and socially responsible companies that are users of genetic resources (Laird and Wynberg 2005, 5; 2008). This insight is further corroborated by Padmashree Gehl Sampath (2005, 165), who points to capacity building as a means of improving ABS in practice.

Access to seeds and medicinal plants, conservation and the equitable sharing of benefits, and strengthened innovation—these are all worthy objectives, but they are not necessarily mutually compatible. Without access to the genetic resources, there can be little innovation. Without patents, there may be less incentive for innovation; but patent claims that are too broad may also hamper innovation by stifling access to genetic resources (Safrin 2004).[23] A number of initiatives are emerging to respond to this problem, with the aim of finding options that do not hamper innovation and development within basic research.

Conclusion

There is little doubt that the depletion and loss of biodiversity has generated a broad range of international cooperative responses—most important, the CBD. It is more questionable whether the responses have been effective in terms of implementation. We have seen that the parties involved in the CBD came to the negotiation table with rather different agendas in mind. While the asymmetries that affected these negotiations were largely overcome in the agreement, they nonetheless persist in hindering the implementation phase. The Millennium Ecosystem Assessment pointed to an annual human deprivation of US$250 billion as a consequence of biodiversity loss. This perceived scarcity has yet to make political headlines, though. While this may be one major factor hindering implementation (particularly as it relates to the incongruent valuation of biodiversity between the developing and developed countries), it is surprising from the perspective that phenomena like climate change represent one of several threats to the biodiversity and ecosystems that all human well-being depends on. As acknowledged in a recent European Commission (2008) report, "We are still struggling to find the value of nature." Indeed, quotas and markets have not yet been developed to capture this value. This may take increasing shape as the need for natural genetic resources becomes more paramount. ABS could represent one such system of facilitating this valuation—the one this chapter has dealt with, as it grasps the underlying disagreements that pervade the biodiversity issue area. Other systems include various forms of payment for ecosystem services schemes, such as the UN reduced emissions from deforestation and forest degradation initiative that is emerging under the climate change regime.[24]

The scope for any of these systems to deal effectively with biodiversity conservation and equity, however, remains to be seen. The rapid growth of bilateral free trade agreements is a strong indication that industrialized countries, pushed by strong multinational corporations, have succeeded in advancing their agenda and goals for strengthened intellectual property rights and patent systems.[25] While the normative impact of the ABS argument can be observed in bilateral bioprospecting deals at the corporate level, there remains a lack of knowledge about the scope and actual implementation in terms of behavior as well as the needs of public and private actors in bioprospecting deals across different sectors. This requires additional exploration, as there is an unfortunate inconsistency between the need for a multilateral system and the pitfalls of choosing a one-size-fits-all solution. Moreover, there is still quite some way to go in terms of establishing a legally binding multilateral regulatory framework, which would strengthen the weaker parties in bioprospecting deals. One, albeit weak, proposed balancing measure—the disclosure of origin of genetic resources—has yet to be accepted internationally and integrated into patent legislation in user countries.

The success of a multilateral system is likely to depend on compatible legislation in user and provider countries—in order to balance strengthened intellectual property rights systems. The bilateral implementation provisions so far stipulated in the CBD on benefit sharing have not been helpful in terms of empowering poor countries in bioprospecting deals. Hence, the major asymmetries resurface as stumbling blocks to cooperation on biodiversity loss in the implementation phase.

Since the early phases of CBD negotiations, user countries have nevertheless come a long way in realizing the need for reciprocity through some kind of multilateral system. Furthermore, the ABS issue is one where the G-77 coalition has been able to develop a strong common position with accompanying norms (Najam 2002). Although implementation efforts remain low, the potential and incremental steps of norm diffusion should not be disregarded. It is also interesting to consider the contrafactual situation and surmise what the contents of bioprospecting deals would have looked like in the absence of the CBD. The CBD has certainly raised awareness about ABS issues and engendered norms, which affect what is perceived as legitimate in international transactions with genetic resources. The contrafactual situation regarding conservation efforts is more difficult to judge. The comprehensive scope of the CBD is probably necessary, but as we have seen, it is not sufficient for effective implementation and goal achievement.

Notes

1. Convention on Biological Diversity, article 2. Available at <http://www.cbd.int/convention/articles.shtml?a=cbd-02> (accessed November 16, 2009).

2. See Stokke and Thommessen 2003/2004. Resolution 8/83, twenty-second session of the FAO conference, Rome, 1983. Available at <ftp://ftp.fao.org/ag/cgrfa/Res/C8-83E.pdf> (accessed November 16, 2009).

3. The distinction between the North and South is problematic as there are certainly great differences within these categories, with regard to both the biodiversity endemism and technology levels. Still, this general dichotomy has been persistent in the CBD negotiations and it is useful in shedding light on the asymmetries that affect the parties' behavior.

4. These two reports are available at <http://www.unep.org/Documents.Multilingual/Default.asp?DocumentID=55&ArticleID=163&l=en> (accessed November 16, 2009); <http://www.unep.org/search.asp?q=UNEP+State+of+the+Environment+&cx=007059379654755265211%3Ajkngxjgnyii&cof=forid%3A11&sa.x=0&sa.y=0&sa=go!#938> (accessed November 16, 2009).

5. At the same time, high intensification and input factors as well as monocultures associated with domesticated biodiversity within agriculture are major sources of biodiversity loss through habitat conversion, degradation, and pollution (Bishop et al. 2008).

6. As an illustration, less than 1 percent of flowering plants have been thoroughly investigated for their chemical composition.

7. Furthermore, the diversity of marine species is largely unknown; the deep sea may rival tropical forests in species diversity. Microbial diversity is still also largely unknown, and so is insect diversity (Laird and Wynberg 2005; Haefner 2003; Sheldon and Balick 1995, 105).

8. For a discussion of the differences between distributive and integrative bargaining, see Underdal 1980. For a discussion of commons resources and property rights, see Ostrom 1990.

9. The International Undertaking was the first comprehensive international agreement dealing with plant genetic resources for food and agriculture. It was adopted by the FAO conference in 1983 (resolution 8/83) and revised in 1989. One hundred and thirteen countries have adhered to the International Undertaking.

10. In 1980, in the so-called *Chakrabarty Case*, the U.S. Supreme Court for the first time ruled in favor of patenting naturally occurring biological materials. Since then, patents in biotechnology have become increasingly broad, and the patent criterion for inventions has been lowered (Barton and Berger 2001; Safrin 2004).

11. All but Andorra, Holy See, and the United States have become parties to the treaty.

12. Strategic Plan for the Convention on Biological Diversity, UNEP/CBD/COP6/decision 26, Hague, Netherlands, 2002. "Annex B: Mission, § 11: Parties commit themselves to a more effective and coherent implementation of the three objectives of the Convention, to achieve by 2010 a significant reduction of the current rate of biodiversity loss at the global, regional and national level as a contribution to poverty alleviation and to the benefit of all life on earth." Available at <http://www.cbd.int/decision/cop/?id=7200> (accessed November 16, 2009).

13. Bonn Guidelines on Access to Genetic Resources and Fair and Equitable Sharing of Benefits Arising out of Their Utilization. UNEP/CBD/COP6/decision 24, Hague, Netherlands, 2002. Available at <http://www.cbd.int/decision/cop/?id=7198> (accessed November 16, 2009).

14. Implementation is often defined as actual behavioral change in line with the agreed-on political objectives (van Meter and van Horn 1975).

15. A crucial question, however, is whether this issue-specific power consists of the more enduring *basic game power*, referring to the parties' control over the resources in question, or *negotiation power*, referring to capabilities based on strength in numbers, coalitions, and leadership during the negotiations (Underdal 1997, 17).

16. The interaction between the CBD and the Kyoto Protocol of the UN Framework Convention on Climate Change is a case in point, as the two may give rise to different and conflictive practices and instruments. The regulations of the Kyoto Protocol tend to rank timber production (e.g., through plantations) as a more important concern than the protection of biodiversity. Viewing forests primarily as sinks—or carbon reservoirs—is different from appreciating their value in terms of the full range of the plant and animal species that they accommodate, however. From the perspective of biodiversity conservation, replacing old-growth forests with plantations represents a major threat (see Rosendal 2001a, 2001b).

17. See <http://www.gefweb.org/> (accessed February 6, 2009).

18. Cofinancing is the part of GEF funding that is from sources other than public grants. It is often in the form of private-public partnerships. More information is available at <http://www.gefweb.org/interior_right.aspx?id=50> (accessed November 16, 2009). In effect, from 1991 and 2002, the climate change area received an estimated US$5,000 million in cofinancing compared to US$2,000 million for biodiversity (Pearce 2004). During the period from 2002 to 2006, another US$20 billion was allocated in cofinancing, but only about US$3.17 billion of this has gone to biodiversity. See <http://www.gefweb.org/interior_right.aspx?id=224> (accessed February 6, 2009).

19. Ad Hoc Open-ended Working Group on Review of Implementation of Convention, *Note by the Executive Secretary*, UNEP/CBD/WG-RI/2/INF/4, June 28, 2007. Page 2, § 3 reads: "While unmet conservation financing needs in developing countries are difficult to gauge precisely, indicators are that as much as US$20–US$50 billion per annum will be needed over the next decade." Available at <http://www.cbd.int/doc/meetings/wgri/wgri-02/information/wgri-02-inf-04-en.pdf> (accessed November 16, 2009).

20. Article 1 states that "the objectives of this Treaty are the conservation and sustainable use of plant genetic resources for food and agriculture and the fair and equitable sharing of the benefits arising out of their use, in harmony with the Convention on Biological Diversity, for sustainable agriculture and food security."

21. Prior art pertains to the patent criterion of novelty and the question of whether the patented object was previously known. This is often defined and delimited to the search of whether it has been published in an English-language journal.

22. The CBD parties have unsuccessfully been negotiating a legally binding instrument on ABS since 2002. They aim to agree on such a regime at the conference of the parties in October 2010 in Nagoya, Japan. The negotiation text is, however, still full of brackets, that is, controversial items. With a view to reaching the goals of ABS and conservation, failure may be preferable to a watered-down regime.

23. Regarding the pharmaceutical sector, Médecins Sans Frontières (2001) argues: "In fact, patents may actually hamper medical research in developing countries. Patents are often owned by private companies or research institutions that during the period of protection, put limits on research knowledge. Molecules that could be promising for the treatment of neglected diseases are consequently not easily accessible for research."

24. See <http://www.un-redd.org/UNREDDProgramme/tabid/583/language/en-US/Default.aspx> (accessed November 17, 2009).

25. There is a rather high correlation between industry-dominating countries and key economic interests in the global economy. The United States is dominating the biotechnology industry, accounting for 78 percent of global company revenues, followed by Europe at 14 percent (Ernst and Young 2005).

References

AEWA (Agreement on the Conservation of African-Eurasian Migratory Waterbirds). 1995. June 16. Available at <http://www.unep-aewa.org/> (accessed November 16, 2009).

African Convention on the Conservation of Nature and Natural Resources. 1968. September 15. Available at <http://www.jus.uio.no/treaties/06/6-01/african-conservation-nature.xml> (accessed November 16, 2009).

Agreement on the Conservation of Gorillas and Their Habitats. 2007. October 26. Available at <http://www.cms.int/species/gorillas/index.htm> (accessed November 16, 2009).

Andersen, Regine. 2008. *Governing Agrobiodiversity: Plant Genetics and Developing Countries*. Aldershot, UK: Ashgate.

Andresen, Steinar, and G. Kristin Rosendal. 2007. "The role of the United Nations Environment Programme in the coordination of multilateral environmental agreements." In *International Organizations and Global Environmental Governance*, ed. Frank Biermann, Bernd Siebenhüner, and Anna Schreyögg, 133–150. London: Routledge.

Barrett, Scott. 2003. *Environment and Statecraft: The Strategy of Environmental Treaty Making*. Oxford: Oxford University Press.

Barton, John, and Peter Berger. 2001. "Patenting agriculture." Issues in Science and Technology Online. Available at <http://www.issues.org/17.4/barton.htm> (accessed June 8, 2006).

Barton, John, and Wolfgang E. Siebeck. 1992. "Intellectual property issues for the international agricultural research centres: What are the options?" *Issues in Agriculture* 4, CGIAR Secretariat, The World Bank, Washington DC.

Bishop, Joshua, Sachin Kapila, Frank Hicks, Paul Mitchell, and Francis Vorhies. 2008. *Building Biodiversity Business*. London: Shell International Limited and the International Union for Conservation of Nature.

CBD (Convention on Biological Diversity). 1992. June 5. Available at <http://www.cbd.int/convention/convention.shtml> (accessed March 11, 2008).

Chambers, Jasemine. 2002. "Patent eligibility of biotechnological inventions in the United States, Europe, and Japan: How much patent policy is public policy?" *George Washington International Law Review* 34 (223): 237–239.

Choudry, Aziz. 2005. "Corporate conquest, global geopolitics: Intellectual property rights and bilateral investment agreements." *Seedling* (January): 7–13.

CITES (Convention on International Trade in Endangered Species of Wild Fauna and Flora). 1973. March 3. Available at <http://www.cites.org/> (accessed November 16, 2009).

CMS (Convention on Conservation of Migratory Species). 1979. June 23. Available at <http://www.cms.int/pdf/convtxt/cms_convtxt_english.pdf> (accessed November 11, 2008).

Convention between the United States, Great Britain, Russia, and Japan for the Preservation and Protection of Fur Seals [known above as North Pacific Fur Seal Treaty]. 1911. July 7. Available at <http://fletcher.tufts.edu/multi/sealtreaty.html> (accessed November 16, 2009).

Convention on the Conservation of European Wildlife and Natural Habitats. 1979. September 19. Available at <http://conventions.coe.int/treaty/en/Treaties/Html/104.htm> (accessed November 11, 2009).

Convention on Wetlands of International Importance, especially as Waterfowl Habitat [known above as Ramsar Convention on Wetlands]. 1971. February 2. Available at <http://www.ramsar.org/cda/ramsar/display/main/main.jsp?zn=ramsar&cp=1_4000_0__> (accessed November 11, 2008).

Correa, Carlos M. 1999. "Access to plant genetic resources and intellectual property rights." *Background Study Paper* 8. Rome: FAO Commission on Genetic Resources for Food and Agriculture.

Crespi, Stephen R. 1988. *Patents: A Basic Guide to Patenting in Biotechnology.* Cambridge: Cambridge University Press.

Dutfield, Graham. 2001. "Biotechnology and patents: What are developing countries doing about article 27.3(b)?" *Bridges* 5 (9): 17–18.

Ernst and Young. 2005. *Beyond Borders: A Global Perspective.* New York: Ernst and Young.

EUROBATS (Agreement on the Conservation of Populations of European Bats). 1991. December 4. Available at <http://www.eurobats.org/> (accessed November 16, 2009).

European Commission. 2008. *The Economics of Ecosystems and Biodiversity.* European Communities. Wesseling, Germany: Welzel and Hardt.

FAO (Food and Agriculture Organization). 1998. *The State of the World's Plant Genetic Resources for Food and Agriculture.* Rome: Food and Agriculture Organization.

Fowler, Cary. 2001. "Protecting farmer innovation: The Convention on Biological Diversity and the question of origin. *Jurimetrics* 41 (4): 477–488.

Gehl Sampath, Padmashree. 2005. *Regulating Bioprospecting: Institutions for Drug Research, Access, and Benefit-Sharing.* Tokyo: United Nations University Press.

Gleckman, Harris. 1995. "Transnational corporations' strategic responses to sustainable development." In *Green Globe Yearbook*, ed. Helge Ole Bergesen, Georg Parmann, and Øystein B. Thommessen, 93–106. Oxford: Oxford University Press.

Grajal, Alejandro. 1999. "Biodiversity and the nation state: Regulating access to genetic resources limits biodiversity research in developing countries." *Conservation Biology* 13 (1): 6–10.

Haefner, Burkhard. 2003. "Drugs from the deep: Marine natural products as drug candidates." *Drug Discovery Today* 8 (12): 536–544.

Hendrickx, Frederic, Veit Koester, and Christian Prip. 1993. "Convention on Biological Diversity—access to genetic resources: A legal analysis." *Environmental Policy and Law* 23 (6): 254–255.

Heywood, Vernon H. 1995. *Global Biodiversity Assessment*. Cambridge: Cambridge University Press.

Hutton, Jon, and Barnabas Dickson. 2000. *Endangered Species, Threatened Convention: The Past, Present, and Future of CITES*. London: Earthscan.

International Treaty on Plant Genetic Resources for Food and Agriculture. 2001. November 3. Available at <ftp://ftp.fao.org/docrep/fao/011/i0510e/i0510e.pdf> (accessed November 16, 2009).

Kloppenburg, Jack R. 2004. *First the Seed: The Political Economy of Plant Biotechnology*. Cambridge: Cambridge University Press.

Kloppenburg, Jack R., and Daniel L. Kleinman. 1987. "The plant germplasm controversy." *Bioscience* 37 (3): 190–198.

Koester, Veit. 1997. "The biodiversity convention negotiation process and some comments on the outcome." *Environmental Policy and Law* 27 (3): 175–192.

Laird, Sarah A. 2000. "Benefit-sharing 'best practice' in the pharmaceutical and botanical medicine industries." In *Bioprospecting: From Biodiversity in the South to Medicines in the North*, ed. Hanne Svarstad and Shivcharn S. Dhillion, 89–99. Oslo: Spartacus Forlag.

Laird, Sarah A., and Rachel Wynberg. 2005. The Commercial Use of Biodiversity: An Update on Current Trends in Demand for Access to Genetic Resources and Benefit-Sharing, and Industry Perspectives on ABS Policy and Implementation. UNEP/CBD/WGABS/4/INF/5.

Laird, Sarah A., and Rachel Wynberg. 2008. "Access and benefit sharing in practice: Trends in partnerships across sectors." *Secretariat of the Convention on Biological Diversity. Montreal. Technical Series* 38.

Lanchbery, John. 2006. "The Convention on International Trade in Endangered Species of Wild Fauna and Flora (CITES): Responding to calls for action from other nature conservation regimes." In *Institutional Interaction: Enhancing Cooperation and Preventing Conflicts between International and European Environmental Institutions*, ed. Sebastian Oberthür and Thomas Gehring, 157–180. Cambridge, MA: MIT Press.

Martens, Pim, Jan Rotmans, and Dolf de Groot. 2003. "Biodiversity: Luxury or necessity?" *Global Environmental Change* 13 (2): 75–81.

McNeely, Jeffrey A., Kenton R. Miller, Walter Reid, Russel Mittermeier, and Timothy B. Werner. 1990. *Conserving the World's Biological Diversity*. Washington, DC: World Resources Institute.

Médecins Sans Frontières. 2001. "Fatal imbalance: The crisis in research and development for drugs for neglected diseases." MSF Access to Essential Medicines Campaign and the Drugs for Neglected Diseases Working Group. Available at <http://www.accessmed-msf.org> (accessed June 8, 2006).

Millennium Ecosystem Assessment. 2005. *Ecosystems and Human Well-being: Biodiversity Synthesis*. Washington, DC: World Resources Institute.

Moser, Michael, Crawford Prentice, and Scott Frazier. 1996. A Global Overview of Wetland Loss and Degradation. Available at <http://www.ramsar.org/about/about_wetland_loss.htm> (accessed June 8, 2006).

Myers, Norman. 1996. "The biodiversity crisis and the future of evolution." *Environmentalist* 16: 37–47.

Najam, Adil. 2002. "Unravelling of the Rio bargain." *Politics and the Life Sciences* 21 (2): 46–50.

Oberthür, Sebastian, and Thomas Gehring, eds. 2006. *Institutional Interaction: Enhancing Cooperation and Preventing Conflicts between International and European Environmental Institutions*. Cambridge, MA: MIT Press.

Ostrom, Elinor. 1990. *Governing the Commons: The Evolution of Institutions for Collective Action*. Cambridge: Cambridge University Press.

Pearce, David. 2004. "Environmental market creation: Saviour or oversell?" *Portuguese Economic Journal* 3 (2): 115–144.

Porter, Gareth. 1993. "The United States and the Biodiversity Convention: The case for participation." Paper presented at the International Conference on the Convention on Biological Diversity, Nairobi, January 26–29.

Quammen, David. 1996. *The Song of the Dodo: Island Biogeography in an Age of Extinctions*. New York: Touchstone.

Raustiala, Kal, and David G. Victor. 2004. "The regime complex for plant genetic resources." *International Organization* 58 (2): 277–309.

Reid, Walter V., and Kenton R. Miller. 1989. *Keeping Options Alive: The Scientific Basis for the Conservation of Biodiversity*. Washington, DC: World Resources Institute.

Rosendal, G. Kristin. 1991. *International Conservation of Biological Diversity: The Quest for Effective Solutions*. R: 012–1991. Lysaker: Fridtjof Nansen Institute.

Rosendal, G. Kristin. 2000. *The Convention on Biological Diversity and Developing Countries*. Dordrecht: Kluwer Academic Publishers.

Rosendal, G. Kristin. 2001a. "Impacts of overlapping international regimes: The case of biodiversity." *Global Governance* 7 (1): 95–117.

Rosendal, G. Kristin. 2001b. "Overlapping international regimes: The case of the Intergovernmental Forum on Forests (IFF) between climate change and biodiversity." *International Environmental Agreement: Politics, Law, and Economics* 1 (4): 447–468.

Rosendal, G. Kristin. 2006a "The Convention on Biological Diversity: Tensions with the WTO TRIPS agreement over access to genetic resources and the sharing of benefits." In *Institutional Interaction: Enhancing Cooperation and Preventing Conflicts between International and European Environmental Institutions*, ed. Sebastian Oberthür and Thomas Gehring, 79–102. Cambridge, MA: MIT Press.

Rosendal, G. Kristin. 2006b. "Regulating the use of genetic resources—between international authorities." *European Environment* 16 (5): 265–277.

Safrin, Sabrina. 2004. "Hyperownership in a time of biotechnological promise: The international conflict to control the building blocks of life." *American Journal of International Law* 98: 641–685.

Schei, Peter J. 1997. "Konvensjonen om biologisk mangfold." In *Rio + 5: Norges oppfølging av FN-konferansen om miljø og utvikling*, ed. William M. Lafferty, Oluf S. Langhelle, Pål Mugaas, and Mari Holmboe Ruge, 244–258. Oslo: Tano Aschehoug.

Schroeder, Richard A. 2000. "Beyond distributive justice: Resource extraction and environmental justice in the tropics." In *People, Plants, and Justice*, ed. Charles Zerner, 52–66. New York: Columbia University Press.

Sheldon, Jennie Wood, and Michael J. Balick. 1995. "Ethnobotany and the search for balance between use and conservation." In *Intellectual Property Rights and Biodiversity Conservation*, ed. Timothy M. Swanson, 45–64. Cambridge: Cambridge University Press.

Simpson, David R., Roger A. Sedjo, and John W. Reid. 1996. "Valuing biodiversity: An application to genetic prospecting." *Journal of Political Economy* 104 (1): 163–185.

Stokke, Olav Schram, and Øystein Thommessen, eds. 2003/2004. *Yearbook on International Cooperation on Environment and Development*. London: Earthscan.

Swanson, Timothy, ed. 1995. *Intellectual Property Rights and Biodiversity Conservation*. Cambridge: Cambridge University Press.

ten Kate, Kerry, and Sarah A. Laird. 1999. *The Commercial Use of Biodiversity: Access to Genetic Resources and Benefit-Sharing*. London: Earthscan.

Tolba, Mostafa K. 1998. *Global Environmental Diplomacy: Negotiating Environmental Agreements for the World, 1973–1992*. Cambridge, MA: MIT Press.

TRIPS (Trade-Related Aspects of Intellectual Property Rights). 1994. April 15. Available at <http://www.wto.org/english/thewto_e/whatis_e/tif_e/agrm7_e.htm> (accessed November 16, 2009).

Tvedt, Morten W. 2005. "How will a substantive patent law treaty affect the public domain for genetic resources and biological material?" *Journal of World Intellectual Property* 8 (3): 311–344.

Underdal, Arild. 1980. *The Politics of International Fisheries Management: The Case of the Northeast Atlantic*. Oslo: Universitetsforlaget.

Underdal, Arild. 1997. "Modelling the international climate change negotiations: A non-technical outline of model architecture." Center for International Climate and Energy Research (CICERO) Working Paper 8. Oslo: CICERO.

UNDP (United Nations Development Programme). 2005. *Human Development Report*. New York: United Nations Development Programme.

van Meter, Donald S., and Carl E. van Horn. 1975. "The policy implementation process: A conceptual framework." *Administration and Society* 6 (4): 445–488.

WCED (World Commission on Environment and Development). 1987. *Our Common Future*. Oxford: Oxford University Press.

WHC (Convention concerning the Protection of the World Cultural and Natural Heritage). 1972. November 23. Available at <http://whc.unesco.org/en/convention> (accessed November 16, 2009).

Wilkinson, Clive, ed. 2004. *Status of Coral Reefs of the World: 2004*. Townsville: Australian Institute of Marine Science.

Wilson, Edward O., ed. 1988. *Biodiversity*. Washington, DC: National Academy Press.

Wilson, Edward O. 1992. *The Diversity of Life*. New York: W.W. Norton.

WIPO (World Intellectual Property Organization). 2005. *Intellectual Property and Traditional Knowledge*. Booklet 2, publication 920E. Geneva: World Intellectual Property Organization.

WRI (World Resources Institute). 1993. *Biodiversity Prospecting*. Washington, DC: World Resources Institute.

WRI (World Resources Institute), IUCN (World Conservation Union), UNEP (United Nations Environment Programme) in consultation with the FAO (Food and Agriculture Organization) and UNESCO (United Nations Education, Scientific, and Cultural Organization). 1992. *Global Biodiversity Strategy: Guidelines for Action to Save, Study, and Use Earth's Biotic Wealth Sustainably and Equitably*. Washington, DC: World Resources Institute.

Young, Oran R. 1996. "Institutional linkages in international society: Polar perspectives." *Global Governance* 2 (1): 1–24.

III

Scarcity and Degradation in a Regional Context

5

Transboundary Cooperation to Address Acid Rain: Europe, North America, and East Asia Compared

Miranda A. Schreurs

Acid rain is a serious challenge afflicting many parts of the world, including Europe, North America, and East Asia. Acid rain, the popular term used to describe the deposition of acidic compounds, can result in considerable ecological destruction and affect human health as well. It can raise the acidity levels of soil and water, thereby adversely impacting plant and animal life. Not only can acid rain damage trees, agricultural crops, and aquatic life but it can slowly corrode buildings and damage equipment. It stems from air pollutants, including sulfur dioxide, nitrogen oxides, and volatile organic compounds, that can be deposited at distances both near and far—hundreds and even thousands of miles away—from their original source. As a result, acid rain is not only a local environmental problem but can also become a transboundary environmental issue. The compounds that contribute to acid rain come primarily from the burning of fossil fuels in utility plants, industries, automobiles, ships, and homes (Visgilio and Whitelaw 2007).

As the primary sources of acid rain are not always located in the same states or regions that suffer most from the environmental degradation caused by this acidic deposition, acid rain can be a source of interstate tension. Chinese factories and power plants not only cause acid rain within China, they also impact Japan and the Korean Peninsula. Reports suggest that acid rain from China is even reaching distant North America (Wilkening 2004a). U.S. coal-fired power plants located in the Midwest contribute to acidic deposition not only in New England but also in eastern Canada. In 1987, Ontario's minister of the environment traveled to New York "to persuade Americans that this insidious killer of fish and other aquatic organisms must be swiftly eliminated."[1] Nitrogen oxide emissions from fossil fuel burning and automobile traffic in the more populated and industrialized European states cause problems not only domestically but also in sparsely populated regions like Scandinavia and ecologically sensitive areas like the Alps. Acid rain is in many ways an upwind/downwind environmental problem.

Acid rain has many important features relevant to the themes of this volume: scarcity, conflict, and cooperation. Although acid rain is usually discussed as an environmental pollution problem, it is interesting to view it through an environmental scarcity lens. Acid rain is tied to a wide variety of forms of environmental degradation, ranging from the acidification of lakes and ocean coastlines to damages to forests, crops, infrastructure, and buildings. This means that when acid rain is present, especially in large quantities and sensitive environments, it can lead to a variety of environmental scarcity concerns, including the "lack" of healthy air, soils, lakes, and streams (due to an excessively high acid content).

A particularly large source of acidic pollutants is coal. Some of the regions that have been most heavily affected by acid rain are those where coal is the primary source of fuel. In fact, the term acid rain was first introduced in 1852 by the chemist Roger Angus Smith, who worked in Manchester, one of the most heavily polluted regions of nineteenth-century England. A century later, in the mid-1990s, Manchester was still suffering from its coal-based roots. It was the unlucky recipient of the label "Europe's capital for acid rain." According to a study conducted by a Trinity College team, "Acid rain dissolved buildings in Manchester faster than at any other test site from Donegal to Athens. . . . The city suffered worst because its rain was the most acid."[2] The city, it can be argued, suffers from an environmental scarcity problem.

Acid rain seriously affected many European countries, especially in the 1970s, 1980s, and 1990s, and caused considerable diplomatic conflict among them, with downwind states objecting to the acid rain sent their way. Yet as the extent of acid rain and its damage in an increasingly large part of Europe became clearer, cooperation on acid rain abatement slowly emerged. Today, Europe is arguably the region with the most developed international cooperative framework for combating acid rain and other transboundary air pollution problems: the Convention on Long-range Transboundary Air Pollution (CLRTAP). The CLRTAP is an example that fits well with the thesis of this book that "environmental scarcity" can lead to cooperation. It is also an instance of what is increasingly referred to as an environmental "success story" (Global Atmospheric Pollution Forum 2007; Hareide 2004).

A similar pattern occurred between Canada and the United States. In the 1980s, diplomatic tensions mounted as Canada became increasingly frustrated by the unwillingness of the U.S. government to introduce policies to control sulfur dioxide and nitrogen oxide emissions. Here too it can be maintained that environmental scarcity triggered bilateral conflict. And as in the European case, as the acid rain problem became increasingly serious, especially in the northeastern United States and Canada, growing concerns about acid rain did lead to cooperation. Joint scientific cooperation on acid rain began in the 1970s. After the U.S. Congress passed the Clean Air Act Amendments of 1990, introducing a sulfur dioxide emissions trading system, Canada and the United States formed the Air Quality Agreement of 1991 to control sulfur dioxide and nitrogen oxides, both precursors to acid rain. Since then, the United States and Canada have experienced substantial reductions in their sulfur dioxide and, to a lesser extent, nitrogen oxide emissions. In this case too, environmental scarcity in the form of acidified lakes and soils likely led to international (bilateral) cooperation.

Yet as will be argued in more detail below, cooperation occurred years later in the North American than the European case even though Canada, the United States, and the countries of Europe were all founding members of the CLRTAP. In addition, there were different policy responses in Europe and North America. Europe moved over time to a critical loads approach to the control of pollutants; in the United States, an emissions trading system was introduced. Domestic politics strongly influence when and in what forms cooperation may emerge. The reasons why the North American and European responses to long-range transboundary acid rain differed in terms of both the timing and nature of their policy responses will be discussed below.[3]

A more difficult case for the contention that environmental scarcity will result in international cooperation is East Asia. Acid rain is particularly severe in East Asia and has been a factor in regional tensions. The countries of East Asia have initiated scientific cooperation on acid rain. They have failed, however, to introduce a regional regime for the control of emissions that can result in acid rain. In other words, the nature and depth of the cooperation that has emerged in East Asia differs substantially from that found in the North American and European cases. In East Asia, cooperation is limited to joint scientific studies and specific air pollution control projects. East Asia lacks the kind of formal, international air pollution agreements that exist in Europe and North America.

As these three cases illustrate, there is power to the claim that environmental scarcity can cause conflict as well as foster international cooperation—even in regions where such cooperation might otherwise not be expected. Still, it is too simplistic to suggest that all states and regions will respond in the same way to similar environmental scarcity problems. While it is the case that as acid rain has become increasingly serious it has pushed both pollution-importing and pollution-exporting states into international cooperative arrangements, interstate cooperation in addressing transboundary acid rain has varied significantly in how it has "localized" acid rain science as well as what it asks of participants across Europe, North America, and Asia (Wilkening 2004a). The thesis that environmental scarcity can lead to cooperation tells us little about the form that this cooperation may take, its depth, or the timing of its initiation.

The goal of this chapter is to understand the similarities and differences in how a particular environmental scarcity problem has been addressed in three regions that have all been hit by acid rain. The comparison examines the extent to which acid rain has triggered conflict as well as cooperative problem solving in these different parts of the world, exploring why a cooperative regime to address acid rain emerged in Europe earlier than it did between Canada and the United States, and why there is still no acid rain emissions control regime in East Asia. The chapter begins with a brief historical discussion of acid rain politics in Europe, North America, and East Asia. These case studies draw on the sizable body of extant literature addressing acid rain politics in each of these regions. The purpose of this exercise is therefore not to shed new light on acid rain politics in any one of these regions but rather to consider what these case studies can tell us about acid rain as an environmental scarcity issue, and its link to interstate conflict and cooperation. The comparison takes into consideration political, economic, and environmental asymmetries among the states, and the ways in which these asymmetries have played into regional acid rain responses. The chapter also explores why different policy instruments and measures have been adopted to address acid rain in the three regions.

Acid Rain and Environmental Scarcity

Acid rain interacts with natural environments in a variety of ways. Some regions are more vulnerable to acid rain than others, depending on such factors as soil type, vegetation, altitude, and the level of rainfall. Some soils and water bodies have a higher buffering capacity and can neutralize acid. Others have little buffering capacity, and as a result, when acid rain falls, the pH of the soil and water bodies decline (that is, they get more acidic), affecting plant and aquatic life. Highly acidified lakes are unable to support aquatic life. Highly acidified soils may result in extensive damage to forests and eventual forest death. Coastal and mountainous regions and aquatic environments are particularly susceptible to acidic deposition. In these ways, acid rain can be considered a cause of other forms of environmental scarcity.

In Europe and North America, acid rain impacts have been widespread, and have resulted in both lake death and forest damage. In China, there are also many reports of ecological damage due to acid rain. In contrast, in Korea and Japan where the buffering capacities of soils is higher, there has been more limited damage. A 2003 report by the South Korean government to the UN Forum on Forests stated that soil acidity had been increasing since 1973, but that only 12 percent of sixty-five locations monitored were "sensitive to forest damage" (Republic of Korea 2003). Neither Japan nor South Korea has experienced substantial forest or lake death as a result of acid rain coming from China. This does not mean that there is not substantial concern with acid rain, however. Different from the cases of Europe and North America, where acid rain was initially largely treated as a distinct environmental scarcity issue, in East Asia it is being viewed as one aspect of a larger pollution problem. China's air pollution is threatening long-term environmental quality not only in China but also in Japan and South Korea (Schreurs 2001). As a Reuters article attests: "Smog adds to a string of environmental concerns that experts say originate in China, including acid rain and sandstorms that gain toxicity as they pass over industrial regions."[4]

The European Response to Acid Rain: The CLRTAP Agreement and Its Eight Protocols

Acid rain has been a known phenomenon for well over a century. It was not until the 1960s and into the 1970s, though, that the potential for acidic compounds to travel hundreds of miles in the form of rain, snow, and dry precipitation began to be understood. Somewhat ironically, the problem of the long-range transport of acidic compounds was exacerbated by policies that were introduced in the 1950s and 1960s to address local air pollution. In order to improve ambient air quality in regions close to industrial sites, tall chimney policies were established in numerous countries. This reduced pollution close to the source, but contributed to environmental scarcity problems (a lack of healthy air, water, and soil) over wider and wider areas.

Internationally, the first region to feel the effects of this tall stack policy was Scandinavia. In the 1960s, Svante Oden, a Swedish soil scientist, helped to bring the problem of acid rain to the attention of the Swedish public and policymakers, and eventually the international community. Concern about the impacts of acid rain on Sweden's lakes and forests was an important factor in Sweden's decision to host the UN Conference on the Human Environment in 1972. At this conference, Sweden put the spotlight on the ecological damage that the long-range transport of acid rain appeared to be causing in Scandinavia. Over the course of the next decade, various countries along with the Organization for Economic Cooperation and Development (OECD) introduced research programs into long-range acidic deposition. Sweden's role was pivotal not only in focusing international attention on a "new" scientific problem, long-range air pollution, but also in initiating a global political forum for the discussion of acid rain and other common pool environmental scarcity issues (Clark et al. 2001).

Bringing science into the policy process was not so easy, however. For the better part of the decade, Great Britain, Germany, and Eastern Europe challenged the plausibility of Sweden's claims (Park 1988). And although in 1979 the CLRTAP was formulated among European countries, Canada, and the United States, the initial agreement simply called on countries to reduce their transboundary air pollution as much as was economically feasible, report on their efforts, and conduct joint research. Sweden pushed at this stage to have a requirement for a 30 percent reduction in each country's sulfur dioxide emissions written into the agreement, but this idea was opposed by Germany and the United Kingdom. Strong domestic interests (and especially the coal lobbies) blocked action.

With the formation of CLRTAP, cooperation had been achieved, but its depth was still limited. The Scandinavian countries thus pursued an innovative alternative. They began a voluntary cooperative initiative—a "30 percent club" of states that would agree to unilaterally cut back their sulfur dioxide emissions by 30 percent. They were joined by Canada (which was having its own problems with the United States), France, Switzerland, Austria, the Netherlands, the Soviet Union, and quite remarkably, Germany (Levy 1995).

Germany's participation was crucial, and it signaled the beginning of an important change in environmental policy approaches in Europe's largest economy and manufacturing center. Germany had initially done its best to brush aside Scandinavian complaints about its emissions. Yet by the early 1980s, German politics were becoming increasingly "green." There were several critical factors behind Germany's change of heart. They had to do with more general concerns about growing environmental scarcities. Two particularly significant factors relevant to this chapter were the growing public awareness and concern about acid rain damage to German forests, and the electoral success in 1982 of the Green Party, which for the first time gave the Greens a presence in the federal parliament. By the next year, Chancellor Helmut Kohl's conservative government was introducing legislation to reduce emissions from large combustion sources and then taking this to the European level so that there would be a more level playing field for industry. The end result was the European Community Large Combustion Plant Directive. The change in policy direction in Germany also made it possible to introduce the Helsinki Protocol in 1985 to the CLRTAP, calling for a 30 percent reduction in sulfur dioxide emissions. Noticeably absent from the agreement were Poland, the United Kingdom, and the United States—all big coal states. This was the first protocol to the CLRTAP to propose a concrete emissions reduction target.

European cooperation on acid rain was expanding, but key actors were still only half engaged. The United Kingdom was especially relevant in this regard. It was participating in scientific research initiatives, but opposing concrete emissions reduction targets. Cooperation was thus limited. UK resistance to cooperation on firm emissions reduction targets could be explained by the lack of strong public concern with acid rain and the monopolistic, state-owned Central Electricity Board, with its interest in expanding coal- and nuclear-based utilities.

Much as had occurred in Germany in the early 1980s, a change of policy direction on acid rain began to occur in the United Kingdom in the late 1980s as concerns about environmental scarcities more generally took firmer root in society. Several factors were critical in this process. One was the growing sense of environmental crisis after the discovery of the "ozone hole" over Antarctica by British scientists in 1985 and the Chernobyl nuclear accident. Another and perhaps more important factor was then–Prime Minister Margaret Thatcher's interest in privatizing the Central Electricity Board and breaking the power of the coal lobby. The United Kingdom experienced a dash away from coal to natural gas. Here, specific political interests were essential to tipping the balance in favor of action to address acid rain. The United Kingdom has subsequently ratified all of the protocols to the CLRTAP except the Helsinki Protocol.

While it is easy to explain the interest in the formation of an international regime in states that were importers of pollution, in states like Germany and the United Kingdom that were the sources of pollution, either antipollution advocacy coalitions had to be strengthened or the interests of powerful actors had to be changed to bring about a turn in policy direction in favor of cooperation in addressing acid rain along with the associated environmental scarcity problems.

As mentioned above, German participation can largely be explained by the growing domestic concern with acid rain damage to German forests and the entrance of the Green Party into parliament. This changed the balance of political forces in the German parliament enough to attract the attention of Germany's larger political parties. Britain's change of heart, in the words of Brian Wynne and Peter Simmons (2001, 99), "has been much more due to incidental larger political-economic developments like privatization and the wholesale dash to gas, from coal, than to measures specifically to address acid rain." Thus, in Britain's case broader political economic changes (privatization) had major implications on actor's interests, and this opened the door for greater British cooperation with the CLRTAP regime.

Over time the CLRTAP has become increasingly comprehensive, addressing a range of interrelated pollution problems (acid rain, eutrophication, and ground-level ozone) that contribute to a range of environmental scarcity problems in Europe. The CLRTAP and its protocols now cover sulfur dioxide, nitrogen oxides, volatile organic compounds, heavy metals, and persistent organic pollutants. The twelve countries that became new members of the European Union (EU) between 2004 and 2007 have all been expected to adopt the measures found in the CLRTAP (Andonova 2007, 154). In the case of the new member states, it can be argued that it was not concern with acid rain itself that led to policy changes and regional cooperation but rather the expectations as well as the side payments in the form of redevelopment assistance that came with membership in the EU.

The North American Response to Acid Rain: The U.S. Clean Air Act Amendments of 1990 and the 1991 Air Quality Accord with Canada

In North America, acid rain became a major point of political contention between the United States and Canada. Due to prevailing winds, coal-fired power plants in the United States contributed to acid deposition in Canada, plus the "death" of Canadian lakes and streams. In response to growing scientific concerns about these environmental scarcity issues, the United States and Canada established a bilateral research consultation group in 1978 to exchange scientific information related to acid rain issues, and both were founding parties to the CLRTAP agreement in 1979. In 1980, the Memorandum of Intent concerning Transboundary Air Pollution established cross-border working groups to develop the scientific and technical basis for the development of a bilateral agreement.

Formal negotiations for a bilateral agreement were broken off in 1982. There was no linear progression toward deeper cooperation. Rather, there was an abrupt turn away from cooperative engagement, resulting in a sharp cooling of U.S.-Canadian relations. The change in the political environment under President Ronald Reagan stymied many of the efforts to achieve a bilateral agreement on acid rain that had been initiated under his predecessor, Jimmy Carter. As one of his major goals, Reagan wanted to scale back the federal government in general; in particular, he wanted to reduce environmental regulations. Thus, the Reagan administration pulled the United States out of acid rain negotiations with Canada (Clark and Dickson 2001, 264). Over the course of the next decade, Canada became increasingly "inventive, aggressive, and desperate" in its diplomatic campaign to convince the United States to restrict its sulfur dioxide emissions (Parson et al. 2001, 238). Frustrated in its inability to alter U.S. policy, Canada began to look to the international arena for new avenues to attempt to affect change. Canada became a strong supporter of Sweden's proposed 30 percent club and organized a meeting in 1984 for the club's members (U.S. officials attended as observers). Domestically, the seven eastern Canadian provinces agreed in 1985 on a province-by-province allocation of reductions and cost-sharing arrangements with a goal of substantially reducing sulfur dioxide emissions (Schreurs et al. 2001, 359). Canada joined many European countries in the establishment of domestic emissions reduction targets. Cooperation was achieved, but it was not between upwind and downwind states. Instead, it was largely among the "victim" states—those importing pollution.

It was not only the eastern Canadian provinces that were suffering from the Midwest's pollution; the northeastern U.S. states were suffering as well. They too began an increasingly strong lobbying campaign in Congress. For years there was a congressional deadlock as representatives of the coal-burning states opposed action and those in the northeastern states, being adversely impacted by emissions, pushed for the polluters to pay. With a divided Congress and a lack of presidential leadership on the issue, inaction ensued (Bryner 1995).

Policy change in the United States required a window of opportunity. This occurred with the election of a new president, George H. W. Bush, who wished to establish a better relationship with the environmental community than his predecessor had maintained. The change was also aided by the availability of a relatively new policy idea, which was more amenable to Bush's conservative constituency: an emissions trading system for the reduction of sulfur emissions.[5] The emissions trading system provided an innovative solution to the sticky question of who should pay the cost for cleaning up emissions—those emitting the pollution or those suffering from it (Munton 2007).

The ensuing Clean Air Act Amendments of 1990 set targets for reducing sulfur and nitrogen oxide emissions by 50 percent.[6] The innovative aspect of the legislation was that in contrast to earlier command-and-control approaches to emissions reductions, the emissions trading approach introduced a greater element of flexibility into the implementation process. The Clean Air Act Amendments combined the introduction of regulatory caps on sulfur dioxide and nitrogen oxide emissions with an emissions trading system that placed a price on pollution. Utilities could determine whether they wanted to keep paying the price for their pollution allowances (which rose with time), or take steps to reduce their pollution so that they could retire or sell off their allowances to utilities that needed them for their new or expanding operations. Over time, allowances were retired from the system, raising the price of the remaining permits. This market-based system proved more appealing to many industries than the traditional command-and-control approach, establishing technology standard requirements or mandating similar action by all firms regardless of the capacity or cost of making pollution control improvements. It proved to be a rapid and cost-effective means for reducing emissions, especially in its early years.

The Canada-U.S. Air Quality Agreement signed in 1991 required both countries to take on specific emissions reduction goals. The United States basically enshrined targets in the agreement that it had adopted under the Clean Air Act Amendments of 1990. Canada agreed to reduce sulfur dioxide emissions in its eastern provinces and to adopt nitrogen oxide standards for automobiles equivalent to those introduced in the United States under the 1990 amendments (Parson 2001, 239). The agreement also set an institutional framework for cooperation in addressing other air quality issues, such as particulate matter and ground-level ozone (Vaughn 2001, 216–217).

It did not, however, address directly the environmental scarcity aspects of acid rain, as the focus was on emissions themselves rather than the carrying capacity of sensitive environments. Canadian assessments suggest that the level of acidic deposition remains above the critical load level—that is, it continues to threaten long-term environmental sustainability.[7]

The East Asian Response to Acid Rain: The East Asian Acid Deposition Monitoring Network

East Asia is currently the region most threatened by the environmental scarcity issues associated with acid rain. Yet the acid rain problem is difficult to separate from other air pollution problems because they are so closely intertwined. Today, China is considered one of the most seriously polluted spots on the planet, and much of the pollution there is connected to the burning of its abundant sources of coal. The costs of air pollution and acid rain to China in the forms of lost agricultural yields, damaged crops, and human health (all of which can be defined as environmental scarcity problems) have been enormous. In the mid-1990s, it was estimated that close to 20 percent of the agricultural land in seven provinces was affected by sulfur dioxide emissions and acid rain. The negative health effects of air pollution were estimated to amount to over 4 percent of China's GDP in 1995 (Guttikunda et al. 2004, 73–74). In 2003, the Chinese Institute of Environmental Science and Tsinghua University determined that the cost was approximately 110 billion yuan (US$13.3 billion) annually.[8] The World Bank estimates that air pollution costs China 3.8 percent of its GDP (and water pollution another 2 percent).[9]

China's acid rain and other air pollution problems are causing considerable concern downwind on the Korean Peninsula and in Japan. The South Korean National Institute of Environmental Research reported in 2006 that 37 percent of the sulfur dioxides that trigger acid rain in Korea stem from China. The research institute pointed out that the acid rain is linked to an array of skin diseases, particularly in the elderly and children, disruption of the ecosystem, and corrosion of buildings.[10]

The Chinese media now reports regularly on the pollution spewing from Chinese smokestacks: "Guangzhou Swamped by Acid Rain," "Acid Rain Damages Giant Buddah Statue," "One-third of China Hit by Acid

Rain."[11] A Xinhua News Agency article stated ominously: "Acid rain erodes metal, stone, cement and wood, threatening power lines, railway track, houses and bridges. . . . Soils and water that soak up acid rain transfer the pollution to crops and eventually into people's corneas and respiratory tubes."[12] In 2006, the State Environmental Protection Administration (SEPA) admitted that sulfur dioxide emissions in China had reached 25.49 million tons, making China the world's biggest sulfur dioxide polluter. SEPA Deputy Director General Li Xinmin announced that each ton of sulfur dioxide causes China 20,000 yuan (US$2,500) in economic damages, which suggests a whopping loss of about US$64 billion in 2005.[13]

Although China initially denied that its emissions could cause problems in neighboring states, in the mid-1990s it acknowledged the potential for transboundary acid rain to be an issue (Brettell 2007). Slowly as the extent of the acid rain problem in China began to become harder to ignore and the seriousness of the environmental scarcity problems associated with acid rain began to be understood, cooperative efforts to address the problem emerged. The Acid Deposition Monitoring Network in East Asia (EANET) was created at the initiative and with the financial support of the Japanese Environment Agency in the mid-1990s. Concerned about potential future acid rain damage, the Japanese government launched EANET with the initial goal of standardizing and strengthening monitoring capacities in the region. The Japanese government bolstered its commitment to this endeavor with training programs and the provision of monitoring equipment through its overseas development assistance program (Takahashi 2000). In 1998, the first intergovernmental meeting of EANET was convened and included ten countries. Its membership has since expanded, and there are now thirteen member countries: Cambodia, China, Indonesia, Japan, Lao P.D.R., Malaysia, Mongolia, Myanmar, the Philippines, the Republic of Korea, Russia, Thailand, and Vietnam. These countries, combined, have developed fifty-one monitoring sites. EANET also works to promote public awareness of acid rain, and has begun to provide "useful inputs for decision-making at local, national and regional levels aimed at preventing or reducing adverse impacts on the environment caused by acid deposition."[14]

Paralleling the emergence of EANET, the National Institute of Environmental Research of the Republic of Korea launched joint research among Japan, China, and South Korea on Long-Range Transboundary Air Pollutants in Northeast Asia (LTP). The goals of LTP are to foster scientific exchange related to acid rain in the region (Kim 2007).

The work of the EANET and LTP are regularly recognized in the annual Tripartite Environmental Ministers Meeting (TEMM) held among Japan, China, and Korea. At the ninth TEMM in late 2007, the environment ministers issued a joint communiqué in which they reaffirmed their support for EANET as well as for joint research on LTP. In December 2006, Japanese foreign minister Taro Aso (who later became the prime minister) announced that Japan would provide China with 793 million yen (US$6.82 million) to set up a system for the monitoring of acid rain and yellow sand. At the time Aso stated that "this is a really serious matter. Japan has been contacting China about setting up these systems for a while." The grant is to be used to establish monitoring stations at fifty locations in China.[15]

In sum, since the mid- to late 1990s, countries of the region have begun coordinated scientific monitoring of acid rain and initiated some joint acid rain mitigation projects. So to this extent, the argument that environmental scarcity can trigger international cooperation holds. But cooperation remains limited. International cooperation has been largely constrained to the exchange of information and cooperative scientific monitoring under the EANET's rubric (Brettell 2007; Wilkening 2004b). This is basically where the CLRTAP regime in the late 1970s and early 1980s was in Europe. There is no international agreement or regime in East Asia that addresses the control of emissions or the protection of ecological systems from acid rain. Acid rain control remains a primarily national political matter despite the gravity of the region's atmospheric pollution problems—a mix of acid rain, yellow sand (from desert dust storms), and in recent years, photochemical oxidants (also known as photochemical smog).[16]

A Comparative Examination of Regional Responses to Acid Rain: Scarcity, Country Asymmetries, and Additional Factors

What do these three regional cases tell us about the potential for interstate cooperation in cases of environmental degradation and resource scarcity? Also, how do country differences help to explain cooperation or lack thereof in the case of acid rain? As mentioned in the introductory remarks, the country asymmetries most salient in the acid rain case include the upwind/downwind relationship, the varying effects of acid rain on the respective states, and the economic and political differences among the countries. Additional important factors are the role of information diffusion and domestic politics.

Environmental Scarcity and Conflict

First, in all three regions the discovery of transboundary acid rain resulted in interstate tensions. As Detlef Sprintz and Tapani Vaahtoranta (1994) have contended, the extent to which acid rain damages are "felt" domestically and the costs of taking abatement action will influence a state's likelihood of supporting international cooperation. Downwind victims are more likely to support international cooperation than upwind emitters. As Sprintz and Vaahtoranta theorized, the Scandinavian countries and Canada were those lobbying the hardest for regional cooperation to control acid rain. Japan and South Korea also have urged China to take action to control its emissions.

What needs to happen to encourage upstream polluters to cooperate? Initially, despite the complaints of downwind victim states, exporters of pollution challenged the validity of their neighbors' charges. As interstate tensions mounted, diplomatic efforts were attempted to ease relations. These tended to lead to agreements to establish joint scientific research. The CLRTAP and its first protocol, for example, focused on multilateral scientific research on acid rain and enhancing monitoring capacities. It included no specific emissions reduction goals. Canada and the United States initiated joint scientific research into acid rain within the OECD, and bilaterally, beginning in 1978. They did not agree on emissions reduction measures until over a decade later. Japan launched the EANET in order to promote scientific research into acid rain and develop monitoring capacities across much of East and Southeast Asia. There are still no emissions reduction agreements in the region.

Looking across the three regions a similar pattern is evident: after political tensions emerged due to transboundary acid rain and concomitant environmental scarcity problems, government-backed international scientific cooperation on acid rain science and monitoring was initiated.

Asymmetries and Environmental Cooperation

Second, cooperative problem solving is more likely when downwind and major polluting countries both suffer recognizable ecological damage. This was clearly the case with Germany. Germany became more willing to enter into a cooperative regime once ecological damages related to acid rain were discovered within its own borders. As the popular German newsmagazine *Der Spiegel* summarized the situation: "In 1983 forest death moved the Germans more than AIDS or atomic bombs."[17] Acid rain became a major issue in the 1982 elections that led to the Green

Party's entry into the federal parliament the following year. In 1983, the German government introduced the Large Combustion Plant Ordinance to control emissions from major power plants and signaled its willingness to cooperate with the development of protocols to the CLRTAP.

Similarly, the fact that China began to suffer major agricultural losses from acid rain is a key reason behind why the Chinese government took measures to shut down some of its worst polluting industries and improve automobile fuel efficiency once it became clear that air pollution was having serious health, environmental, and economic consequences. It is also a reason why China was willing to discuss transboundary acid rain as a regional problem. With strong evidence that acid rain was not only an ecological but also an economic burden for China, China's leaders have become increasingly open to joint action. This has made it possible for the environmental ministers of China, Japan, and South Korea to put acid rain on the agenda of their annual trilateral environmental ministerial meetings.

Yet signs of environmental damage in an upwind state do not always suffice to stimulate international cooperation. Science can be framed by actors in ways that they find convenient. In the United Kingdom, despite a strong campaign from Norway, Sweden, and other international actors to pressure the United Kingdom into action as well as a Nature Conservancy Council report in 1984 that characterized the effects of acid rain in Britain on plants and animals as alarming, the government called for more scientific study into the problem (Park 1988). The scientific evidence was framed as insufficient to take action. It was not until a political decision to shift the domestic energy structure from its heavy reliance on coal to the greater use of natural gas, which led to a sharp drop in emissions, that the United Kingdom had a change of heart and agreed to act to control emissions contributing to acid rain.

In the United States too ecological damages related to acid rain were a clear problem for the New England states. Signs of serious acid rain damage in the New England states, however, did not result in congressional action. There was thus an upwind/downwind problem not only between the United States and Canada but also between the midwestern coal states and the eastern and northern U.S. states. In the end, the New England states felt compelled to join the Canadian provinces in lobbying for a change in the U.S. federal air pollution policies. Legislative change (the Clean Air Act Amendments of 1990), as noted earlier, only occurred when Bush was elected as president. Bush needed a major environmental success story to present an environmentally friendlier image to the U.S. public than had his predecessor, Reagan. Indications of the emergence of increasingly severe environmental scarcity problems due to acid rain in both eastern Canada and the northeastern United States did not immediately result in political change. Rather, the political climate in the United States had to change before policy change became possible.

Environmental Scarcity, Public Mobilization, and Policy Change

Third, public and nongovernmental organization (NGO) protest along with media coverage helped to raise awareness of acid rain and the related environmental scarcity issues, and place pressure on policymakers to act. As acid rain came to be understood as a problem by the affected publics, the pressure mounted on policymakers to take action. In both Europe and North America in the 1980s, the lack of government action on acid rain led environmental activists to take dramatic protest actions that were intended to attract media attention and pressure governments to pass legislation controlling emissions. In 1984, Greenpeace activists simultaneously hung banners from smokestacks in eight European countries. The banner hung in Germany read "To the Government: Living Beings Require Air"; in Austria it proclaimed "First the Forests Die, then the People"; and in the Netherlands, it simply stated "Stop Acid Rain." In 1987, a Greenpeace activist scaled Mount Rushmore to protest U.S. government inaction (Durland 1987). Many of the protests led to arrests, as when U.S. environmentalists followed their European counterpart's lead: "Five Greenpeace members protesting acid rain pollution came down from smokestacks in Ohio, Indiana and Arizona . . . into the arms of waiting police officers."[18] With growing evidence of damage from acid rain pollution, NGOs, environmentally oriented opposition parties, the media, and the public started to demand political action. In Europe, this domestic pressure was an important factor in Germany's decision to support the CLRTAP protocols. In the United States, it was a crucial reason for the Clean Air Act Amendments of 1990. Equally noteworthy is that the limited coverage of acid rain in Europe and North America in the media today means that the issue is really only on the agenda of a small community of bureaucrats, scientists, and NGOs. The issue has been superseded by other environmental and environmental scarcity concerns. This is not the case in East Asia, though, where acid rain is still newsworthy.

Social unrest related to pollution is currently on the rise in China, with a reported fifty-one thousand pollution-related protests in 2005, including many tied to poor air quality (Economy 2007). In China's case, it is less clear that there has been protest specifically tied to acid rain than to a wide range of intertwined environmental and air quality conditions. Edwin Lau, the assistant director of the Hong Kong branch of Friends of the Earth, stated that because of the damages it causes to marine environments and agricultural soils, acid rain could lead to social instability in areas hard-hit by acid rain, as Chinese crop growers and fish farmers faced with growing environmental scarcity increasingly struggle to earn a living in a worsening environment.[19]

International Networks and the Diffusion of Ideas

Fourth, the sharing across regions of knowledge and experience through the exchange of ideas and information among as well as between regional communities of scientists, bureaucrats, NGOs, and other experts (what is sometimes called an epistemic community) has helped to spread awareness of acid rain, and has been a persuasive factor behind regional cooperation (Haas 1990; Alm 2000; Wilkening 2004b). Domestic proponents of action often use the examples of other regions to bolster their claims for the need for action. The diffusion of policy innovations from first movers to other states has facilitated efforts at cross-national policy harmonization (Busch, Jörgens, and Tews 2005; Schreurs et al. 2001) and regime formation. This occurred to some extent, for example, in the early stages of the CLRTAP regime. In the absence of a regional agreement, national efforts were undertaken to reduce sulfur dioxide emissions. Many states followed Scandinavia's lead in adopting a 30 percent sulfur dioxide emissions reduction goal (the 30 percent club described above). This diffusion among like-minded states facilitated the development of subsequent regional agreements, since it was easier for these countries to agree to establish a protocol given that they already had policies in place domestically that would meet the protocol's aims. Another example where diffusion made a difference was the establishment of EANET, which was clearly based on European models of scientific cooperation and cooperative monitoring.

Regional Differentiation despite the International Exchange of Ideas

Fifth, despite international scientific networks and policy diffusion across countries, the three regions have embraced distinctly different regional approaches to addressing acid rain control. In the European case, the RAINS-Europe model was developed and now underpins the CLRTAP and its protocols. In the 1990s, Europe shifted away from its initial approach to the control of emissions, which established a common emissions reduction target (e.g., a 30 percent reduction target for sulfur dioxide and nitrogen oxide emissions) for all states regardless of whether they were major polluters or not, and whether or not they were major sufferers of acid rain damage. This approach elicited considerable resistance from some states. Europe has since moved to a system based on a critical loads approach—the RAINS-Europe model developed by scientists at the International Institute for Applied Systems Analysis (IIASA). This model determines "appropriate" national policies based on a variety of input factors, including scientific assessments of acid rain effects on different ecological systems, different abatement options, the level of pollution contributions, and the cost of abatement programs in different countries and regions (IIASA 2008). The primary goal of this approach is to develop policies that will prevent acidic deposition that exceeds the limit beyond which ecological damage is likely to occur. This critical load differs depending on the ecological system. It is interesting to note that the RAINS-Europe model has been used for the development of a RAINS-Asia model (IIASA n.d.). The RAINS-Asia model is not linked to any abatement regimes, however. It is purely informational at this point.

In the U.S. case, industrial opposition to environmental regulations (opposition that was partly related to cost concerns) also prompted major changes in policy direction. In contrast with the critical loads approach adopted in the CLRTAP regime, the United States introduced a cap-and-trade system for the control of sulfur dioxide emissions.[20] This system does not address critical loads and thus the needs of different ecological systems; rather, it stresses the reduction of emissions at the least cost and with great flexibility for emitters in how they can best reduce their emissions. Still, the United States appears to be increasingly interested in the critical loads approach.[21]

Different Levels of Cooperation: International Scientific Cooperation versus the Formation of Binding International Regimes

Finally, a willingness to take some form of domestic action does not equate to a willingness to enter into a formal binding international agreement or regime to address environmental scarcity concerns. The East Asia case points to the difficulty of establishing international legal cooperation under particular circumstances. There are several important obstacles to acid rain regime formation in East Asia. Many of these have to do with political and economic asymmetries. Others are tied to historical memories.

There is a large economic disparity in the region. In contrast with both Europe and North America, where at the time of the CLRTAP launch the main states in question were all reasonably and relatively economically well off, in East Asia there are major economic disparities between China and the other developing countries of Southeast Asia, on the one hand, and Japan and South Korea, on the other. Cooperation—in the form of regime formation—is easier to achieve among countries of relatively similar economic levels. When economic disparities are large, questions of how cleanup costs will be paid become major stumbling blocks. Supporting this view, Nagoya University professor Tsuneo Takeuchi suggested that as China's wealth grows and that of Japan's declines (at least in relative terms), cooperation could become easier.[22] Beyond this, while China is rapidly growing, it is still in many ways a developing country that is heavily dependent on coal for energy. Acid rain abatement is but one of many environmental and development challenges facing China's government leaders. The downwind recipients of acid rain from China—Japan and South Korea—are also being asked to foot a substantial part of the bill for air pollution control in China through technology and financial transfers. There are limits to the willingness of either country to make such transfers on a large scale. Thus, while there are numerous specific cooperative initiatives for acid rain monitoring and abatement, cooperation tends to be on a project-by-project basis. Cost factors play a major role in the calculations of states regarding cooperation. It may prove more effective for South Korea and Japan to find ways to assist China to build up its own capacity to control acid rain and other air pollutants as well as address environmental security problems than to enter into a formal regime with China on acid rain. For different reasons, China, Japan, and South Korea may all favor more nuanced and informal agreements to formal, legally based regimes (Yoon 2006). None of the three are eager to enter into an agreement that might prove too costly.

In China, although environmental scarcity problems (the scarcity of clean air, healthy soil, and water) tied to acid rain from domestic sources are severe, the eastern downwind regions of China are not in a position to join Japan and South Korea in calling for action, as happened in North America when the eastern U.S. states joined the eastern Canadian provinces in lobbying for policy change in the United States. The more closed and restrictive political system in China limits such transnational synergies. It is important to note that there are very few region-specific multilateral regimes of any kind binding the northeast Asian states, despite a long list of environmental scarcity issues. Political and historical factors continue to restrict deeper cooperation.

Whereas in some regions formal regimes regulating the precursors to acid rain can be found (Europe and North America), in others (East Asia), cooperation is limited to project-based initiatives and information exchange. Deeper levels of cooperation—those that require states to give up a degree of sovereignty, agree to joint regulatory action, or accept some extent of international monitoring—require some level of mutual trust. Where there is no sense of regionalism or else interconnectivity among the states of a region, cooperation can still occur, but it tends to be relatively limited. While there may be scientific exchange, there is often no effort at regulatory harmonization or legally based regime formation (Lee 2007).

Regime formation is also easier among countries that have experience with other forms of cooperation. In Europe, where environmental cooperation is highly advanced and many environmental laws are now being formulated in Brussels rather than by the member states, cooperation may have been easier to realize than even between two close allies like the United States and Canada. In East Asia, where there are few legally based and region-specific international agreements of any kind, the lack of a formal agreement to address acid rain is but one of countless examples of a serious regional pollution problem for which no international regime has been forged. There is still considerable historical distrust among these countries that has inhibited regime formation, and thus there is also little regional experience to draw on. In sum, regional regime formation appears more likely among states of relatively similar economic levels and political systems.

Conclusion

Comparing the North American, European, and East Asian examples shows that there were important differences in the way that the acid rain issue came on to the political agenda, and in the timing and nature of the policy responses. At issue is not just whether scarcity leads to cooperation but also how long it takes before cooperation is achieved and what cooperation entails. Domestic political factors, including the interests of powerful political leaders as well as the influence of pro- and antipolicy change coalitions, can greatly influence whether a country's policy makers will embrace the warnings of scientists about environmental scarcity concerns. Beyond this, country asymmetries (from upwind/downwind dynamics to differences in economic level and political system form) play pivotal roles in the evolution of cooperation over acid rain.

Cooperation on acid rain and air pollution abatement is highly advanced in Europe, and there are signs that while problems remain, the CLRTAP and its eight protocols have been quite effective. Cooperation has expanded over time. A number of directives have been introduced to control volatile organic compounds, reduce the sulfur content of gasoline, and limit the sulfur emissions from ships. A revised Large Combustion Plant Directive, instituted in 2001, extended the range of plants covered, and tightened emissions limit values for sulfur dioxides, nitrogen oxides, and dust (Schreurs 2007). The EU is also actively promoting policies to address climate change that will have positive benefits for acid rain abatement. The emissions trading system can be expected to enhance energy efficiency. Policies to promote renewable energies and improve automobile fuel efficiency will have positive air quality benefits as well (Andonova 2007).

The North American bilateral Air Quality Agreement depends heavily on the national acid rain abatement policies of the United States and Canada. In addition, it sets out areas for joint scientific and technical research into emissions and ecosystem monitoring, atmospheric modeling, environmental effects, control technologies, and economic measures, among other measures (Munton 2007, 185). Substantial reductions in sulfur dioxide and nitrogen oxide emissions were achieved by the Clean Air Act Amendments of 1990. Cooperation in terms of joint scientific and technical research has been less successful (Munton 2007, 185). Furthermore, while some North American states and provinces are following the European lead in promoting renewable energies and introducing emissions trading systems for carbon dioxide, North America lags behind Europe in its efforts to systematically address air pollution as an integrated set of problems (including dust, soot, smog, acid rain, climate change, volatile organic compounds, and the like). This suggests that there is still considerable room for cross-societal learning.

The case of East Asia is even more dramatic and pressing. Pollution has reached chronic proportions in East Asia. Regional efforts to monitor emissions and build scientific understanding are expanding, and have had some positive capacity-building effects. Yet there is potential and especially the need for considerably deeper cooperation in the scientific understanding of acid rain in the region, its monitoring, and the development of technological, social, and political options for reducing emissions. At the second Trilateral Ministerial Meeting of Science and Technology Cooperation in March 2007, senior-level administrators and scientific experts from Japan, China, and Korea discussed the potential and need for greater cooperation to address "common serious issues: exhaustible petroleum resources, domestic and trans-boundary environmental problems such as air pollution, acid rain and global warming caused by usage of fossil fuels."[23] Environmental scarcity concerns are growing, and so too is regional dialogue.

Environmental cooperation can be a vehicle to promote dialogue among states that have strained security relations (the IIASA was set up in Vienna at the height of the Cold War to encourage joint scientific research into environmental problems among the Soviet bloc states, Europe, and North America). While it may be unrealistic to expect an East Asian regime establishing pollution control limitations, East Asia could benefit from greater and deeper regional cooperation in integrated air pollution research, monitoring, and technological development. Greater cooperation that took an integrated approach to addressing the regions' dust and sandstorms, acid rain, and climate change could help facilitate greater exchange among East Asian countries, and lay the foundations for a more peaceful and cooperative regional politics (Hyun and Schreurs 2007).

In conclusion, the relationships among environmental scarcity, conflict, and cooperation are complex, and affected by both domestic political factors and international structures and processes. Yet some patterns are evident. Environmental scarcities associated with acid rain have contributed to disputes between pollution exporters and importers. In the context of public pressure for change and a sufficiently strong supporting coalition, environmental scarcity has been an important factor in opening the door to international cooperation. The depth of international cooperation, however, differs considerably and appears to be more limited when the asymmetries between states are greater.

Notes

1. Nelson Bryant, "Outdoors; Canada steps up its acid-rain campaign," *New York Times*, February 1, 1987, available at http://www.nytimes.com/1987/02/01/sports/outdoors-canada-steps-up-its-acid-rain-campaign.html?pagewanted=1 (accessed October 8, 2008).
2. "Acid Rain Is Dissolving Manchester," *New Scientist* 1982, June 17, 1995, 13.

3. Here, only the U.S. and Canadian cases within North America will be considered even though transboundary acid rain is certainly an issue along the Mexico-U.S. border; for a history of this, see Liverman and O'Brian 2001.

4. Chisa Fujioka, "Smog Smothers Japan, Experts Point to China," *Reuters*, August 24, 2007, available at http://www.reuters.com/article/idUST21446820070824 (accessed April 17, 2010).

5. Discussion with Daniel J. Dudek, Chief Economist, Environmental Defense Fund, September 23, 2005. Dudek was instrumental in developing the sulfur emissions trading concept and met with President George H. W. Bush about it.

6. The U.S. sulfur emissions trading system became a model for the European Carbon Emissions Trading Scheme, which was launched in 2005.

7. Canada-U.S. Air Quality Agreement Progress Report, 2008, available at <http://www.ec.gc.ca/cleanair-airpur/caol/canus/report/2008CanUs/eng/s3_eng.cfm#s3_4_2> (accessed October 8, 2008).

8. "Acid Rain Costs China Annual Loss of 110 Billion Yuan," *People's Daily*, October 11, 2003, available at <http://english.peopledaily.com.cn/200310/11/eng20031011_125803.shtml> (accessed October 8, 2008).

9. "World Bank: China's Air Pollution Costs 3.8% of GDP," *China Economic Review*, November 20, 2007, available at <http://www.chinaeconomicreview.com/dailybriefing/2007_11_20/World_Bank:_Chinas_air_pollution_costs_38_of_GDP.html> (accessed November 27, 2009).

10. Ministry of the Environment, Republic of Korea, "37% of Sulfur Dioxide Comes from China," news release, June 28, 2006, available at <http://eng.me.go.kr/docs/news/press_view.html?topmenu=B&cat=510&seq=337&mcode=&page=7> (accessed November 27, 2009).

11. Chen Hong, "Guangzhou Swamped by Acid Rain," *China Daily*, March 28, 2008, 4; Huang Zhiling and Zheng Degang, "Acid Rain Damages Statue," *China Daily*, April 19, 2005, 3; Li Liu, "One-third of China Hit by Acid Rain," *China Daily*, August 28, 2006, 2.

12. "China Suffers Severe Acid Rain Contamination," *Xinhua News Agency*, September 22, 2006, available at <http://news.xinhuanet.com/english/2006-09/22/content_5126089.htm> (accessed April 17, 2010).

13. Mo Hong'e, ed. "Acid Rain Poses Great Challenge to Goal of Well-off Society," Xinhua News Agency, August 3, 2006, available at <http://news.xinhuanet.com/english/2006-08/03/content_4913667.htm> (accessed April 17, 2010).

14. EANET Web site, available at <http://www.eanet.cc/> (accessed April 17, 2010).

15. Agence France Press, "Japan to Offer Aid to Monitor Acid Rain and Yellow Sand in China," December 9, 2006, available at <http://www.terradaily.com/reports/Japan_To_Offer_Aid_To_Monitor_Acid_Rain_And_Yellow_Sand_In_China_999.html> (accessed April 17, 2010).

16. Discussion with Dr. Tsuneo Takeuchi, professor of environmental studies, Nagoya University, Japan, December 26, 2008.

17. Jochen Bölsche, "Die Rächer der Entlaubten," *Der Spiegel*, March 29, 2008, available at <http://einestages.spiegel.de/static/topicbumbackground/1676/die_raecher_der_entlaubten.html> (accessed October 8, 2008).

18. "5 Held in Acid Rain Protests," *New York Times*, February 13, 1982, available at <http://www.nytimes.com/1982/02/13/us/5-held-in-acid-rain-protests.html> (accessed November 27, 2009).

19. Agence France Press, "Acid Rain Threatening Food Chain," *China Daily*, August 10, 2006, available at <http://www.chinadaily.com.cn/china/2006-08/07/content_658379_2.htm> (accessed April 17, 2010).

20. Discussion with Dudek. It is interesting to note that there have also been small-scale experiments with sulfur emissions trading in China. This is due to the efforts of the U.S. group Environment Defense, which has established operations there.

21. "Critical Loads and Exceedances," Canada-U.S. Air Quality Agreement progress report (2008), available at <http://www.ec.gc.ca/cleanair-airpur/caol/canus/report/2008CanUs/eng/s3_eng.cfm> (accessed April 17, 2010).

22. Discussion with Takeuchi.

23. Ministry of Education, Culture, Sports, Science, and Technology of Japan, "Summary Report of Workshop for the Trilateral Science and Technology Cooperation" (March 5–6, 2007).

References

Alm, Leslier R. 2000. *Crossing Borders, Crossing Boundaries: The Role of Scientists in the U.S. Acid Rain Debate*. Milton Park, UK: Greenwood Publishing Group.

Andonova, Liliana B. 2007. "Acid rain in a wider Europe: The post-Communist transition and the future European acid rain policies." In *Acid in the Environment: Lessons Learned and Future Prospects*, ed. Gerald R. Visgilio and Diana M. Whitelaw, 151–174. New York: Springer.

Brettell, Anna. 2007. "Security, energy, and the environment: The atmospheric link." In *The Environmental Dimension of Asian Security: Conflict and Cooperation over Energy, Resources, and Pollution*, ed. In-taek Hyun and Miranda A. Schreurs, 89–114. Washington, DC: United States Institute of Peace.

Bryner, Gary. 1995. *Blue Skies, Green Politics: The 1990 Clean Air Act and Its Implementation*. Washington, DC: Congressional Quarterly Press.

Busch, Per-Oluf, Helge Jörgens, and Kerstin Tews. 2005. "The global diffusion of regulatory instruments: The making of a new international environmental regime." *Annals of the American Academy of Political and Social Science* 598 (1): 146–167.

Canada-U.S. Air Quality Agreement Progress Report. 2008. "Critical loads and exceedances." Available at <http://www.ec.gc.ca/cleanair-airpur/caol/canus/report/2008CanUs/eng/s3_eng.cfm#s3_4_2> (accessed October 8, 2008).

Clark, William C., and Nancy M. Dickson. 2001. "Civic science: America's encounter with global environmental risks." In *Learning to Manage Global Environmental Risks: A Comparative History of Social Responses to Climate Change, Ozone Depletion, and Acid Rain*, Vol. 1, ed. William C. Clark, Jill Jäger, Josee van Eijndhoven, and Nancy M. Dickson, 259–294. Cambridge, MA: MIT Press.

Clark, William C., Jill Jäger, Jeannine Cavender-Bares, and Nancy M. Dickson. 2001. "Acid rain, ozone depletion, and climate change: An historical overview." In *Learning to Manage Global Environmental Risk: A Comparative History of Social Responses to Climate Change, Ozone Depletion, and Acid Rain*, Vol. 1, ed. William C. Clark, Jill Jäger, Josee van Eijndhoven, and Nancy M. Dickson, 21–56. Cambridge, MA: MIT Press.

Convention on Long-Range Transboundary Air Pollution. 1979. November 13. Available at <http://unece.org/env/lrtap/full%20text/1979.CLRTAP.e.pdf> (accessed October 8, 2008).

Durland, Steven. 1987. "Witness the guerrilla theater of Greenpeace." *High Performance* 40: 30–35.

EANET (Acid Deposition Monitoring Network in East Asia). Available at <http://www.eanet.cc/eanet/brief.html> (accessed October 8, 2008).

Economy, Elizabeth. 2007. "The great leap backward?" *Foreign Affairs* 86 (5): 38–60.

Global Atmospheric Pollution Forum. 2007. "Tackling regional, hemispheric, and global air pollution: The potential role of the UNECE Convention on Long-range Transboundary Air Pollution." United Nations background document. Sixth Ministerial Conference, Environment for Europe, Belgrade, Serbia, October 10–12, ECE/BELGRADE.CONF/2007/INF/8.

Guttikunda, Sarath K., Todd M. Johnson, Feng Liu, and Jitendra J. Shah. 2004. "Programs to control air pollution and acid rain." In *Urbanization, Energy, and Air Pollution in China: The Challenges Ahead (Proceedings of a Symposium)*, ed. Chinese Academy of Engineering, Chinese Academy of Sciences, National Academy of Engineering, and National Research Council, 73–94. Washington, DC: National Academies Press.

Haas, Peter. 1990. *Saving the Mediterranean*. New York: Columbia University Press.

Hareide, Knut A. 2004. "Reducing air pollution in Europe: The success of the LRTAP Convention." Available at <http://www.unece.org/env/documents/2004/eb/Ministerial%20Statements/Norway.pdf> (accessed October 8, 2008).

Helsinki Protocol on the Reduction of Sulphur Emissions or Their Transboundary Fluxes by at Least 30 Percent [known above as Helsinki Protocol]. 1985. July 8. Available at <http://unece.org/env/lrtap/full%20text/1985.Sulphur.e.pdf> (accessed October 8, 2008).

Hyun, In-taek, and Miranda A. Schreurs. eds. 2007. *The Environmental Dimension of Asian Security: Conflict and Cooperation in Energy, Resources, and Pollution*. Washington, DC: United States Institute of Peace Press.

IIASA (International Institute for Advanced Systems Analysis). 2008. "RAINS and its role in negotiations." Cleaner Air for a Cleaner Future: Controlling Transboundary Air Pollution. Available at <http://www.iiasa.ac.at/Admin/INF/OPT/Summer98/negotiations.htm> (accessed October 8, 2008).

IIASA (International Institute for Advanced Systems Analysis). n.d. "RAINS-Asia." Available at <http://www.iiasa.ac.at/~heyes/docs/rains.asia.html> (accessed October 8, 2008).

Kim, Inkyoung. 2007. "Environmental cooperation of northeast Asia: transboundary air pollution." *International Relations of the Asia-Pacific* 7 (3): 439–462. Available at <http://irap.oxfordjournals.org/cgi/content/full/lcm008v1> (accessed October 8, 2008).

Lee, Guen. 2007. "A regional environmental security complex approach for East Asia." In *The Environmental Dimension of Asian Security: Conflict and Cooperation over Energy, Resources, and Pollution*, ed. In-taek Hyun and Miranda A. Schreurs, 23–40. Washington, DC: United States Institute of Peace Press.

Levy, Mark. 1995. "International co-operation to combat acid rain." In *Green Globe Yearbook of International Co-operation on Environment and Development*, ed. Helge Ole Bergesen, Georg Parmann, and Øystein B. Thommessen, 59–68. Oxford: Oxford University Press.

Liverman, Diana, and Karin O'Brian. 2001. "Southern skies: The perception and management of global environmental risks in Mexico." In *Learning in the Management of Global Environmental Risk: A Comparative History of Social Responses to Climate Change, Ozone Depletion, and Acid Rain*, Vol. 1, ed. William C. Clark, Jill Jäger, Josee van Eijndhoven, and Nancy M. Dickson, 213–234. Cambridge, MA: MIT Press.

Ministry of Education, Culture, Sports, Science, and Technology of Japan. 2007. "Summary report of workshop for the Trilateral Science and Technology Cooperation." March 5–6. Available at <http://www.mext.go.jp/a_menu/kagaku/kokusai/bunsyo/07030817/001.htm> (accessed October 8, 2008).

Munton, Don. 2007. "Acid rain politics in North America: Conflict to cooperation to collusion." In *Acid in the Environment: Lessons Learned and Future Prospects,* ed. Gerald R. Visgilio and Diana M. Whitelaw, 175–202. New York: Springer.

Park, Christopher C. 1988. *Acid Rain: Rhetoric and Reality*. London: Routledge.

Parson, Edward A., with Rodney Dobell, Adam Fenech, Don Munton, and Heather Smith. 2001. "Leading while keeping in step: Management of global atmospheric issues in Canada." In *Learning in the Management of Global Environmental Risk: A Comparative History of Social Responses to Climate Change, Ozone Depletion, and Acid Rain*, Vol. 1, ed. William C. Clark, Jill Jäger, Josee van Eijndhoven, and Nancy M. Dickson, 235–258. Cambridge, MA: MIT Press.

Republic of Korea. 2003. "National reports to the third session of the United Nations Forum on Forests." Available at <http://www.un.org/esa/forests/pdf/national_reports/unff3/south_korea.pdf> (accessed November 27, 2009).

Schreurs, Miranda A. 2001. "Shifting priorities and the internationalization of environmental risk management in Japan." In *Learning in the Management of Global Environmental Risk: A Comparative History of Social Responses to Climate Change, Ozone Depletion, and Acid Rain*, Vol. 1, ed. William C. Clark, Jill Jäger, Josee van Eijndhoven, and Nancy M. Dickson, 191–212. Cambridge, MA: MIT Press.

Schreurs, Miranda A. 2007. "The politics of acid rain in Europe." In *Acid in the Environment: Lessons Learned and Future Prospects*, ed. Gerald R. Visgilio and Diana M. Whitelaw, 119–150. New York: Springer.

Schreurs, Miranda A., William C. Clark, Nancy M. Dickson, and Jill Jäger. 2001. "Issue attention, framing, and actors: An analysis of patterns across arenas." In *Learning in the Management of Global Environmental Risk: A Comparative History of Social Responses to Climate Change, Ozone Depletion, and Acid Rain*, Vol. 1, ed. William C. Clark, Jill Jäger, Josee van Eijndhoven, and Nancy M. Dickson, 349–364. Cambridge, MA: MIT Press.

Sprintz, Detlef, and Tapani Vaahtoranta. 1994. "The interest-based explanation of international environmental policy." *International Organization* 48 (1): 77–105.

Takahashi, Wakana. 2000. "Formation of an East Asian regime for acid rain control: The perspective of comparative regionalism." *International Review for Environmental Strategies* 1 (1): 97–117.

Vaughn, Jacqueline Switzer. 2001. *Environmental Politics: Domestic and Global Dimensions*. 3rd ed. New York: Bedford Books/St. Martin's Press.

Visgilio, Gerald R., and Diana M. Whitelaw. 2007. *Acid in the Environment: Lessons Learned and Future Prospects*. New York: Springer.

Wilkening, Kenneth E. 2004a. "Localizing universal science: Acid rain science and policy in Europe, North America, and East Asia." In *Science and Politics in the International Environment*, ed. Neil Harrison and Gary Bryner, 209–240. Lanham, MD: Rowman and Littlefield.

Wilkening, Kenneth E. 2004b. *Acid Rain Science and Politics in Japan: A History of Knowledge and Action toward Sustainability*. Cambridge, MA: MIT Press.

Wynne, Brian, and Peter Simmons, with Claire Waterton, Peter Hughes, and Simon Shackley. 2001. "Institutional cultures and the management of global environmental risks in the United Kingdom." In *Learning in the Management of Global Environmental Risk: A Comparative History of Social Responses to Climate Change, Ozone Depletion, and Acid Rain*, Vol. 1, ed. William C. Clark, Jill Jäger, Josee van Eijndhoven, and Nancy M. Dickson, 93–115. Cambridge, MA: MIT Press.

Yoon, Esook. 2006. "South Korean environmental foreign policy." *Asia-Pacific Review* 13 (2): 74–96.

6

Conflict and Cooperation in the Mediterranean: MAP from 1975 to Today

Gabriela Kütting

The Mediterranean region combines many challenges, both politically and environmentally. The Mediterranean Action Plan (MAP), designed for all riparian states, was instituted to enable the Mediterranean states to come together to address environmental concerns pertaining to the Mediterranean Sea.[1] Yet it has been of limited effectiveness in bringing about policy-relevant results. It is a highly applicable case study for this volume because of its unusual pattern of cooperation, which has changed substantially throughout the history of the agreement.

When academics talk about the effectiveness of MAP, it is often assumed that MAP is a model of cooperation in a region of conflict where Arabs and Israelis, Greeks and Turks, and the former Yugoslav republics have to coexist, even as a variety of civil wars contribute to instability. This is partially true, but not in the sense of functionalist cooperation nor in terms of combined policy goals. MAP has always been assumed to be a functionalist agreement in which environmental cooperation spilled over into other policy fields and would lead to more confidence-building measures in the region. Such a view cannot be sustained at a closer look, as this chapter will demonstrate. Nevertheless, MAP does show several quite novel ways in which cooperative arrangements take place. These are different from the traditional institutional arrangements we find in the global environmental politics literature, mainly because they are normative in quality rather than policy driven. In other words, the main focus of MAP's raison d'être is not to set common policy goals but instead to spread common norms for the region, which will then find their way into national legislation. Thus, MAP is not a case study that can highlight simple causal connections but rather illustrates the complexities of encouraging states to engage in environmental cooperation in a region that is not as environmentally aware as most industrialized states. An institution focused on norm building is vitally important even if it does not spawn policy goals. It fosters environmental awareness and consciousness, which will in turn lead to environmental policymaking and enable capacity building (see below). In a politically sensitive environment with scarce resources, this is already an ambitious enterprise.

When MAP was first negotiated, its primary motivation was to clean up oil pollution in the Mediterranean Sea. In fact, the "scarcity" of clean seawater was not an issue of conflict or dispute per se; rather, it was an opportunity for cooperation from the outset. Another motivation for MAP was the lack of knowledge about pollutant pathways, and what pollutants were the most pervasive and damaging—again, the scarcity of a clean environment was not an issue threatening conflict, and MAP enabled cooperation here as well. So environmental degradation followed by cooperation and conflict in the region were separate, nonrelated events. Conflict in the region was of a general political nature, though, making it difficult for close links to be forged, which are important for effective policy action. Additional hurdles to effective policy action also had to do with asymmetries such as North-South issues—the financial considerations of poorer versus more solvent member states, but also newly independent ex-colonies that did not want to enter into new forms of dependence through environmental regulation (a type of ideological asymmetry).

Finally, the initial research in MAP suggested that while pollution is a serious issue in the Mediterranean region, most of it is of a nature that does not necessitate joint action as such. As mentioned before, oil pollution and cooperation on how to deal with oil pollution was one of the primary motivations for the establishment of MAP. This was followed by concerns about land-based pollution entering the sea and creating a bathtub effect, diluting the pollution in this semienclosed sea and affecting all the riparian states. Later research did not confirm this assumption and indicated that most pollution remained fairly local. Although this seemed to suggest that international cooperation was not needed, and urgency in the form of coordinated action was therefore reduced, there was still a transboundary effect and lack of common knowledge on how to deal with the problems. So what draws the Mediterranean countries together is the fact that they are facing the same environmental problems rather than a mutual dependence on each other for solutions. Therefore, while cooperation is still motivated by joint goals, the sort of pollution encountered in this region naturally affects the type of solutions developed for environmental degradation, and explains why MAP has evolved to be a normative as opposed to a policy-driven agreement, and why cooperation continues to focus on normative issues and knowledge generation. Capacity building has become an additional goal, as is described below. Both points illustrate how perceived scarcity leads to cooperation, be it normative or policy driven. Looking at the history of MAP and the most urgent environmental threats at different times of its existence will explain this further. In 1995 substantial reforms took place, both in MAP (referred to as MAP II) and the Mediterranean cooperative environment in general, making that period a natural organizing point for discussion.

MAP 1975–1995

MAP is the first and most successful of the United Nations Environment Programme's (UNEP) regional seas programs, which were devised following a series of conferences and meetings by UN agencies and Mediterranean governments on the state of pollution in the sea as well as in reaction to the Stockholm Conference on the Human Environment in 1972 (Raftopoulos 1993). MAP was thus created in 1975 at an interministerial conference, and its legal framework, the Barcelona Convention, was established in 1976 simultaneously with the first two protocols. The Barcelona Convention forms the legal part of MAP and is complemented by a research component (MEDPOL), policy planning programs (the Blue Plan and the Priority Actions Programme [PAP]), and financial/institutional arrangements, covered by UNEP.[2]

In particular, the Mediterranean Sea in the 1970s was a region plagued by substantial pollution, yet no body of knowledge existed that could point to the origins and structures of the problem. It was chosen by UNEP as its flagship plan due to the severe state of the pollution. Thus, the scarcity of clean seawater motivated the institutional organization of MAP, which is based on a three-pronged approach: to research the origins of pollution, take common action in the form of legal agreements, and try to deal with problems of development and their impact on the environment in the policy planning program. The different levels were clearly intended to interact and feed each other at the plan's inception. In practice, however, this has not happened to the extent envisaged and necessary. The lack of interaction may be related to the compartmentalization of MAP as compared to other agreements, but there are many other indicators (e.g., an overall lack of enthusiasm vis-à-vis concrete action, a lack of financial support, and a lack of environmental awareness) that make it impossible to determine the significance of this compartmentalized MAP structure. The various political conflicts in the region and financial disparities as well as a general lack of environmental values were strong contributory factors to MAP's limited form of cooperation.

The institutional arrangement is mainly characterized by two developments: MAP has been extremely successful in building up scientific cooperation along with a scientific body of knowledge about the pollution situation in the Mediterranean Sea, published by UNEP. In addition, MAP has also been successful in constructing a socioeconomic record of environmental and development trends in the region. But this epistemic cooperation was not complemented by politically significant cooperation in a number of areas, such as the political atmosphere between member states, financial considerations, and the role of environmental values.

By 1995, on MAP's twentieth anniversary, the plan had achieved quite a difficult goal by initiating and keeping up a process for more than two decades in a region traditionally hostile to cooperation. In the past ten years, it has moved away from its legal components, focusing on policy priority plans.

I will first describe the four components, discussed above, and then provide an analysis of MAP, highlighting central issues and concerns. Having already explained the motivation for MAP's institutional apparatus, the following sections will elaborate on the manner in which this cooperation has evolved and the problems that have limited MAP's cooperative efforts—such as the political conflicts and economic asymmetries. Since MAP has a complex structure and its components interact, I will clearly separate the four components for descriptive and analytic purposes. Yet this analytic separation does not mean that the components are not interconnected.

The Barcelona Convention and Its Protocols, and the New Convention
The original Barcelona Convention consists of five protocols on:

a. dumping from ships and aircraft (1976)[3]
b. cooperation in combating pollution by oil and other harmful substances in cases of emergency (1976)[4]
c. pollution from land-based sources (1980)[5]
d. specially protected areas (1982)[6]
e. pollution resulting from offshore activities (1994)[7]

While the first three protocols were clearly in response to what was perceived to be the most urgent threats in the region as discernible with common sense but also with the first results of Mediterranean scientific cooperation (the land-based source or LBS protocol listed above), the latter two are a mixture of political motivations and the outcome of scientific cooperation. None of these protocols pursue stringent aims; rather, they set up mechanisms for cooperation without asking for specific commitments from member states. The LBS protocol does set policy aims, yet reproduces these aims from existing international legislation in international treaties.

In 1995, the convention was replaced by a new, amended version, renamed the Convention for the Protection of the Marine Environment and the Coastal Region of the Mediterranean, which takes direct account of the recommendations of the Rio Conference on the Environment and Development in 1992. In addition, the protocol on specially protected areas was replaced by one on biodiversity, and a protocol on hazardous waste was likewise added.[8] With the adoption of MAP II, a second phase in the MAP process was launched, changing its classical pollutant-centered policy approach to an integrative strategy of environmental protection and sustainable development.[9] This is a fundamental change in the approach to cooperation in the region. Parallel commitments to the environmental protection and quality-of-life improvements in the Mediterranean region effectively describe the overall new goal of MAP as environmentally sustainable socioeconomic development. At the same time, MAP II signaled a recognition that lasting environmental protection needs should be included in all socioeconomic policies. MAP therefore strives to ensure the sustainable management of natural marine and land resources, and to integrate the environment into social and economic development and land-use policies. In short, post-1995, MAP has moved away from the traditional convention-protocol legal approach to more community-based approaches. It has effectively closed the door on policy-relevant pollution targets as traditionally found in international environmental treaties and has moved to a normative approach in which norm dissemination takes priority.

MEDPOL

MEDPOL is the scientific component of MAP. When the Mediterranean Sea problem became institutionalized in the 1970s, there was practically no marine science base in the region, and the aim to develop scientific capabilities, especially in the South, was a fundamental aspect of the institution. Peter Haas (1989, 1990) provides data on the distribution of the various marine science disciplines in the Mediterranean region. The science base that did exist or was being developed needed to be standardized so as to use the same measurement techniques as well as to develop a compatible system of analysis throughout the region. One of the benefits for southern MEDPOL participants was the establishment of scientific facilities by UNEP. In a nutshell, MAP has not performed well in making linkages between creating scientific research and transforming it into political action. A substantial part of its early success, however, can be attributed to the establishment of a scientific base in the region, and scientific cooperation works extremely well. When MAP is described as an effective institution, this refers to the success of scientific cooperation rather than political or policy achievements.

Policy Planning

MAP's policy planning component was established in 1977, after two years of negotiations pertaining to the shape it should take. Its scope goes beyond marine pollution issues, extending to regional development and environmental issues in general.

In the planning stages, there were disagreements between the member states as to what a policy planning program should look like. It was generally accepted that the state of the marine environment could not merely be improved by studying individual emissions sources but instead that issues of social and economic planning were directly related to the problem. A two-pronged approach was chosen to deal with the prevailing necessities: a study-oriented program and an action program (Raftopoulos 1993, 22). The policy action program was designed for immediate action, but also for the implementation of the findings of the study plan (the Blue Plan). The two programs were therefore meant to be complementary. This reflects a long-term approach to data collection and knowledge creation concerning an underresearched area.

Moreover, several policy-oriented conferences highlighted policy priority issues. The Genoa Declaration, which focuses on coastal management, was adopted at the Fourth Ordinary Meeting of the Contracting Parties in 1985.[10] It lists ten priority targets for the period 1985–1995. These are: (1) the establishment of reception facilities for ship discharges; (2) sewage plants for all cities of more than a hundred thousand inhabitants and appropriate treatment plants for all towns with over ten thousand inhabitants; (3) environmental impact assessment for all new activities; (4) cooperation on the environmental safety of maritime traffic;

(5) the protection of endangered marine species; (6) concrete measures on pollution reduction; (7) the identification and protection of at least a hundred coastal historic sites of common interest as well as (8) fifty new marine and coastal sites of Mediterranean interest; (9) effective measures on soil erosion, forest fires, and desertification; and (10) a substantial reduction in air pollution.[11] Again, all these policy priority issues were shared goals for the region rather than ones that needed to be achieved through direct cooperation. So their genesis can be reduced back to the successful knowledge generation within MAP as opposed to its successful political cooperation.

These policy recommendations have been much cited, but were not incorporated into the legal or applied component of MAP. Likewise, the 1990 Nicosia Charter on Euro-Mediterranean Cooperation is based on policy recommendations.[12] It brings together the Commission of the European Communities and twelve MAP member states, based on the idea of involving the World Bank and the European Investment Bank as sponsors for a Mediterranean sustainable development program. The aims of the charter are to: confer managerial and financial autonomy to the appropriate environmental institutions; integrate environmental management into socioeconomic development; establish an integrated legislative and regulatory framework; implement environmental impact assessment; adopt economic and fiscal strategies as environmental policy measures; control population growth in the coastal regions; and speed up the completion of the protocol on offshore installations. In addition, by 1993, the coastal zones needing protection were to be designated; ten million inhabitants were to be connected to sewage plants; twenty-five controlled waste dumps for industrial products were to be created, and twenty ports were to be equipped with reception facilities for dirty ballast waters. The Nicosia Charter is an effective attempt to achieve policy targets through bypassing conventional channels of action and involving outside funding agencies, but, it is again based on the same models of cooperation as above: scientific or knowledge generating cooperation, but no policy-driven cross-boundary cooperation.

The Blue Plan was endorsed in 1977, and is a comprehensive study of the economic activities in the Mediterranean Basin and their effect on the environment. It is a collection of data of social and economic trends, which are used as scenarios for future development. The plan is supposed to offer an idea of future trends and act as an indicator for long-term planning. For example, and considering the agrofood sector, the Blue Plan looks at food dependency, the fragility of natural resources, the availability of water resources and usage trends, fertilizer use and effects, demand trends, potential produce shifts, and constraints on the growth of the sector. The agrofood sector is modeled in light of these scenarios (Grenon and Batisse 1989).

In 1977, PAP was also agreed on, with the aim of immediate action in demonstration and pilot projects (Raftopoulos 1993, 27). It started late because of problems in establishing the institutional framework, disagreements about the content, and financial problems. PAP's first aim is not the targeting of environmental degradation but rather a "contribution to the reduction of existing socioeconomic inequalities among Mediterranean states" (Chircop 1992, 23). The areas of concern are integrated planning and the management of coastal zones, aquaculture, the rehabilitation and reconstruction of historic sites, water resources development for islands and isolated coastal areas, and land-use planning in earthquake zones. Although these priority areas were identified, there was no clear idea as to what exactly should be done and where the funding should come from. The United Nations Development Programme offered its support, but a project plan only came into being in 1981 with the financial help of some member states, which had specific programs in mind.

Another policy program is the Coastal Area Management Programs (CAMPs).[13] This scheme was launched in 1989, and provides financial help for member states that plan to develop ecologically sensitive coastal areas (Lempert and Farnsworth 1994, 116). Host states apply for funding for a specific project, which then needs to be approved by MAP officials. The four existing CAMPs are Izmir Bay in Turkey, the island of Rhodes in Greece, Kastela Bay in Croatia, and a project on the Syrian coast.

Altogether, the policy planning programs are good tools to identify some of the pressing issues of the Mediterranean region and a way to attract outside funding to particular subregions or problems. Yet these are not examples of cross-boundary cooperation. Rather, international institutions are used for particular national or local goals. Therefore policy planning programs, through MAP, do not have much influence on international cooperation as such.

Financial and Institutional Arrangements

Originally MAP was administered by the UNEP Regional Seas Office in Geneva. When the Regional Seas Programme increased its number of projects, a special coordinating unit for the Mediterranean was set up in 1979 and then moved to Athens in 1982. Also, a Mediterranean Trust

Fund was established; it relied on the contributions of Mediterranean states as UNEP resources had to be distributed among an increasing number of regional programs. The states involved paid, or were supposed to pay, amounts proportional to their overall UN schedules, which left France with a bill of 48 percent of the overall government contributions (Haas 1990, 125). Before the trust was set up in 1979, MAP was exclusively funded by UNEP, but with the trust individual states were directly responsible for the funding of MAP.

The trust's establishment resulted in the usual late payments and a permanent cash crisis for MAP, thus detracting from its effectiveness. Although UNEP continued to channel resources into the MEDPOL program, policy programs and regional activity centers ran on an extremely low budget. The late payment of members' contributions is a chronic and recurrent problem. For example, by October 1979, thirteen countries and the European Communities had not paid their annual contributions to the Mediterranean Trust Fund, while four members had effected partial payments. Only one country, Greece, had paid all its dues (UNEP/IG.18/4 1979). The figures available for 1995 showed that about one-third of all trust contributions were not expected to be current by December 31, 1995 (UNEP (OCA)/MED.IG.5/16 1995, 38).[14] These financial asymmetries made the effective functioning of MAP difficult.

The resources available are administered through the UNEP/MAP coordinating unit in Athens. Although there are regular meetings of various committees composed of national delegates, which make financial and administrative decisions, the routine running of MAP and the routine allocation of resources are still both done by UNEP via the coordinating unit.

The consequences of bureaucratic layering for MAP are that first of all, the use of the coordinating unit leads to a time lag in deciding on a project and the money becoming available for it. Second, it also means that UNEP remains the driving power behind MAP. Seeing the quibbles and disinterest of the member states in MEDPOL and the policy planning component, this strong presence can be seen as an advantage. Nevertheless, UNEP and especially the coordinating unit are not independent actors since they are composed of individuals from the member states. While this could lead to doubts over MAP's neutrality in allocating project funds, in practice it has not been a problem.

There is definitely a bias in favor of funding MEDPOL projects over PAP, but it is difficult to explain why, except that scientific research was seen as more neutral and less threatening to individual states than PAP research. Moreover, MEDPOL projects do not necessitate concrete policy action whereas PAPs do. It is also perhaps revealing that the main failure of MAP—namely, the missing coordination between MEDPOL and the legal and policy planning components—is exactly the area where UNEP initiatives would not be useful.

The Main MAP Issues

During the 1970s, in the early stages of MAP, immediate action concentrated on the most visible and aesthetically disturbing pollutant: oil. The Mediterranean Sea was clearly ailing at that time, and the need for urgent action was apparent to even a layperson. Little was known about the actual state of pollution in the Mediterranean and therefore the emphasis was placed on knowledge generation in MAP, both with MEDPOL and policy planning—with the idea that knowledge generation would lead to political cooperation.

Tar balls were found on beaches and in the open sea, with visible and tangible effects on human and livestock health. Since the Mediterranean Sea is one of the busiest oil-shipping areas in the world proportionally (20 percent of global oil shipments travel through the Mediterranean Sea), the origins of oil pollution were also assumed to be clear. This explains the focus on oil pollution in the first two protocols of the Barcelona Convention. Ignorance concerning the origins of pollution had to be researched, though, and this process was initiated with the establishment of MEDPOL.

As indicated above, the structure of MAP is designed to research the origins of pollution, take common action in the form of legal agreements, and try to deal with problems of development and their impact on the environment in the policy planning programs. In short, MAP tackles the issue of regional marine pollution on three levels: the generation of knowledge, a commitment to legal aid, and the observation of socioeconomic trends. These different levels were meant to interact and complement each other, especially the research component, by feeding into the policy planning and legal components. As noted above, however, this did not happen in practice to the extent envisaged and necessary.

MAP operates on the basis of an enabling and awareness-raising agreement. It also operates with basically no concept of temporal dimensions, either at the institutional or the environmental level. The emphasis is laid on cooperation and the achievement of a consensus rather than a concrete deadline by which targets have to be achieved. Thus, it puts political capacity building at the top of the agenda as opposed to environmental improvement. The implication of this is that there is a strong awareness of political, financial, and ideological asymmetries between member states, which capacity building could help alleviate.

MAP 1995 to Today

As mentioned earlier, the Barcelona Convention was amended in 1995. As a consequence, its protocols were also substantially revised.[15] The main motivation for the amendments was to express a commitment to the principles of Agenda 21 and other developments arising out of the United Nations Conference on Environment and Development (the "Rio Summit" of 1992). This was followed in 1996 with the establishment of the Mediterranean Commission on Sustainable Development, which was a commitment to develop a regional strategy for sustainable development in the Mediterranean region. This whole process is referred to as MAP II. In effect, it moved the process away from one of legal commitments to one of norm creation. The policy aims of MAP II are thus even fuzzier and less focused than the first phase. At the same time, the politico-environmental process in the region also became much more multilateral.

In 1995, other developments outside the environmental field occurred. The Euro-Mediterranean Partnership (EMP) was established—a partnership between the European Union (EU) and the Mediterranean states.[16] The aim of this collaboration is congruent with general EU goals: economic development, confidence building, and political and economic integration. Like MAP, the EMP is an attempt to foster cooperation in the region through joint programs, building civil society, and funding policy initiatives. While the partnership covers a multitude of issue areas, it excludes environmental concerns. Given that the EU is the single most important environmental actor in the area, this exclusion is deliberate, as this subject area is already covered by MAP. The EMP (also somewhat confusingly referred to as the Barcelona process) somehow suffers from the same fate as MAP, because it is an extremely successful enabling arrangement, but has not spawned much in the shape of concrete projects involving actors that would not normally cooperate. As Peter Liotta (2003, 41) puts it, "While the central concerns of the Barcelona Declaration of 1995 reasonably acknowledge the linkages among cultural, political and economic factors in the Euro-Mediterranean, little multilateral agreement has been reached that will solve the wider negative consequences of collapsing extended security." Here again, we can point to economic and ideological asymmetries as factors hindering policy action from moving beyond capacity building. Arab Hoballah (2006, 167), in contrast, sees the achievements of the Mediterranean Commission for Sustainable Development as "strengthen(ing) the regional commons and regional consensus for sustainable development in the Mediterranean through promotion of the concept at regional and national levels, the participatory approach, the involvement of civil society and NGOs, through capacity development and its learning process approach." Thus, the impact and role of these initiatives are seen as mainly capacity building rather than presenting specific results.

The post-1995 era in Mediterranean environmental politics has not seen any substantial new proposals, projects, or initiatives but has instead focused on consolidating the work of the early years, making it compatible with the new framework aimed at sustainable development. Most progress is achieved at the national rather than the regional level. In fact, if we look at several policy areas that are most pertinent to scarcity situations, a clear picture emerges of a regional institution being used to develop regionwide indicators, but operating mostly at the national level. This again illustrates how MAP is about norm creation as the main mode of cooperation in its second phase.

May Massoud, Mark Scrimshaw, and John Lester (2003, 878) conducted an excellent study of water use and wastewater management in MAP. In the Mediterranean region, nearly three-quarters of all water is used in the agricultural sector. Projections for the future based on agricultural expansion of the region calculate a 32 to 55 percent rise in water use for 2010 and 2025, respectively. In an arid climate with limited water resources in which some areas are already dependent on water deliveries by ship even today, this is a clear scarcity scenario—although again, one where scarcity is perceived at the local level rather than as a cross-boundary issue. Likewise, in a region where wastewater treatment is more often than not in its rudimentary stages, the environmental consequences do not project a favorable picture. Water issues are addressed at the national level in the Mediterranean region. Yet a major problem facing policymakers and the users of water and/or wastewater is a lack of universal standards on wastewater treatment and methods of application. Standard setting at the regional level therefore can play an important role at the national level. In MAP, this was addressed with the Genoa Declaration of 1985, as part of the policy programs, but has not been taken further since. This is another example of the significance of the norm creation and capacity building necessary to develop the political will as well as environmental value base to prepare for joint policies in the future. The EU and World Bank were instrumental in funding projects that linked a substantial number of Mediterranean cities to a sewerage system in the 1990s, but no efforts have been undertaken for rural areas. Outside of private initiatives and voluntary contributions, MAP does not have the financial scope to introduce such policies. So although the Mediterranean has become a multi-institution environment in the 1990s and early twenty-first century, this has not altered the predominantly national outlook of policies for the above reasons.

This finding is confirmed by the most visible MAP component in the early twenty-first century: CAMPs. The pressure on coastal areas has continuously escalated in the Mediterranean region, both in the North and South. Rapid urbanization means a shift in social and cultural patterns, but first and foremost it has an impact on scarce resources such as shifting patterns of food security and water supply. Hence, urbanization is a socioeconomic phenomenon yet has wide repercussions reaching to the security level. Coastal management in the form of special CAMPs organized through MAP and also the integrated coastal management initiative by the EMP have made the Mediterranean region one of the most advanced in the world regarding coastal management. These projects are carried out at the national, subnational, and local levels. Below is an overview of the most well-known projects from the good practices guidelines for integrated coastal area management in the Mediterranean (UNEP, MAP, and PAP 2001):

- *Ulixes 21* has a more or less *supranational character*; it is managed by an NGO and focuses on tourism.
- The *National Master Plan for the Coast of Israel* and *CAMPs Syria* represent initiatives at a *national level*, where the emphasis is placed on integrated planning, institutional reforms, and investment planning.
- The *Strymonikos Project* in Greece refers to a bay, and focuses on the abiotic and biotic characteristics of the marine and terrestrial environment along with a monitoring system of the coastal environment to stress concrete actions.
- *SDAGE* in France refers to a regional scale, presents a unique approach, and concentrates on pollution issues and water resources management. The project links *integrated river basin* and *coastal area management* at a watershed level.
- The *Posidonia Project* concerns a *network* of various coastal urban centers including Greece (Athens), Italy (Naples, Palermo, and Taranto),

and Spain (Barcelona). Networking activities can provide significant benefits, particularly when they are developed among regions encompassing different countries.

- *CAMPs Malta* and the situation of the island complex of *Cyclades* in Greece provide a good opportunity to study not only the complexity of development/environment problems on islands but also to consider alternative options for development, with the aim of promoting concerted actions between islands with common characteristics/problems (i.e., small islands).
- *Venice Lagoon* is an interesting project at a *local level.* It represents a particular case, not only with ecological and historical significance, but with economic significance as well due to tourism development.
- *Kastela Bay CAMPs* in Croatia is another example of local initiatives for integrated coastal area management (ICAM). The emphasis in this case is placed on pollution and urban development issues.
- In *CAMPs Sfax* in Tunisia, the attention is placed on integrated planning, incorporating various interventions with respect to water resources and solid waste management, the monitoring of pollution, and so on.
- The *Belek and Cirali Project* in Turkey deals with sites that are significant from an ecological perspective, such as nesting areas for sea turtles. Belek is an area where large-scale tourism development has taken place, thereby affecting the environment and landscape, while Cirali is an area where traditional activities like agriculture are still carried out.

While one of the projects, Posidonia, reaches across national boundaries, it performs a purely networking function and is limited to the EU states. None of the other projects involve international cooperation. Cooperation is therefore limited to participation in training activities and the joint institutions that have been set up, but does not extend to joint policy initiatives or any other combined activities. This then begs the question of how effective such an enabling or capacity-building arrangement actually is when it comes to preventing conflict or enhancing security. In other words, is capacity-building cooperation a laudable goal in itself or only ever a first step? Does cooperation also have to be policy driven, or can it be equally beneficial with a capacity-based and/or normative approach?

The Effectiveness of Capacity-Building Cooperation

It seems that MAP (and the EMP) takes a two-pronged approach. On the one hand, it has the political aim of increasing cooperation, capacity building, and environmental knowledge in order to lay the political foundations for policy-focused cooperation at a later stage. On the other hand, it has the ecological aim of preventing environmental degradation and controlling pollution. Within certain parameters, MAP is successful at the former and less successful at the latter. One could thus argue that capacity building in the Mediterranean is politically effective but ecologically or environmentally unsuccessful.

When the Barcelona Convention and MAP process came into being in the mid-1970s, as a response to the sea's pollution, no connection was made between the various conflicts in the Mediterranean region and environmental degradation. Rather, a politically tense environment was seen as hampering environmental cooperation. The Arab-Israeli conflict, the Greek-Turkish/Cyprus tensions, postcolonial woes, and general North-South and East-West strains were the main issues in the region. This was juxtaposed with coastal development problems that were considered purely local. Still, the Mediterranean Sea was perceived to be a big "paddling pool," or bathtub, where pollution was diluted to affect everybody given the semi-enclosed nature of the sea. So environmental concerns were not seen as giving rise to conflict, but political conflict was viewed as preventing environmental cooperation. The emphasis on capacity building and trust generation in the MAP process is thus explainable.

When the first results of the established scientific research base indicated that pollution in the Mediterranean Sea tended to remain fairly local and did not generate a bathtub effect, the urgency for common and coordinated action was also not confirmed. This seems to be one of the main reasons why there is no perceived urgency for joint environmental action. Yet the early success of scientific cooperation has given MAP a reputation as a strong environmental agreement. Making a distinction between scientific cooperation and effective environmental policy cooperation explains the discrepancy between this assessment by some analysts and the environmental realities.

With the rise of the environmental security literature (Homer-Dixon 1999; Bächler 1999; Dabelko 2001), the perception of the environment and security changed in both the academy and the region. For example, the Arab or Palestinian-Israeli conflict is often described in terms of water. Water management and the flow of international rivers in the region has become a political and environmental issue, albeit one that is not the subject of MAP II or the EMP, for that matter. Although the socioeconomic and ecological situation of the Mediterranean Basin could be portrayed in human, environmental, and ecological security terms, this has not happened—at least not from within the region. Such a conceptualization could have brought out the need for cooperation or joint action, but the Blue Plan policy component has firmly embedded its research in the sustainability discourse. These two discourses are ideologically compatible, but have not positioned themselves as such and seem to coexist in isolation of each other (Kütting 2007). Here is perhaps the main reason why MAP II does not necessarily emphasize scarcity, cooperation, and conflict as key issues. The sustainability discourse (or at least the part of the discourse driven by states and international institutions) is firmly embedded in a technological-determinist framework that sees technological innovation and increased economic cooperation in a liberalizing economy as the way forward, generating greener economic growth that makes environmental protection affordable (Gillespie 2001; Redclift 1987). And it is exactly in the establishment and extension of the Mediterranean Free Trade Zone that most cooperation in the region has been achieved in any policy sector. This indicates a model of cooperation that is not institutionally based but instead ideologically based.

In the 1990s, the concept of the Mediterranean environment had shifted from one that focused on the actual body of water—the Mediterranean Sea—to one that mainly concentrated on the coastal region. This can be explained by the fact that pollution that takes its origin from sea-based sources has been dealt with through the Barcelona Convention framework, but also by the realization that most pollution emanates from land-based sources, and that land-based pollution has to be addressed through socioeconomic development mechanisms and sustainable development practices. These socioeconomic mechanisms are where MAP II's focus continues to lie in the twenty-first century. But this development also makes cooperation in the region somewhat less traditional than the usual case studies of cooperation and conflict—because of the nonlinear connection between the two. The Blue Plan studies of the PAP in MAP and MAP II (Grenon and Batisse 1989; Benoit and Comeau 2005) clearly indicate that the Mediterranean region is characterized by commonalities: socioeconomic trends in the region are congruent, and demonstrate that there is a common identity for the region or at least the southern Mediterranean. The northern Mediterranean EU states also have strong commonalities, which will be discussed below. Again, such data are aimed at capacity building as well as facilitating cooperation that would not otherwise occur.

In many ways, MAP II has transformed itself into an enabling agreement that fosters cooperation of a different kind, bringing together environment and economic development as compatible goals. In a region where environmental protection is still viewed with suspicion and as an obstacle to economic development, stressing the congruence between the two is a necessary strategy and aimed at overcoming the asymmetries in the region. Nevertheless, it then also ties the hands of MAP II as it does not have the resources to invest in large-scale projects stimulating cooperation and sustainable economic development. Map II opens the door for economic cooperation, however, and essentially gives the green light to economic expansion. Admittedly, the Blue Plan remains critical of aspects of trade liberalization (Benoit and Comeau 2005, 186), but overall it drives the "more sustainability through technology-oriented growth" argument for which the sustainable development discourse has been criticized.

In other words, the cooperation that has taken place under MAP II is not driven by environmental criteria. Although MAP as an institution is firmly committed to environmental improvement, its main mission is not to facilitate cooperation between member states but instead to get particular projects off the ground—projects that are mostly within a state rather than cross-boundary. MAP II promotes interstate cooperation through its commitment to the principles of sustainable development, which are interpreted in a liberal/neoliberal framework in this context. So by promoting liberal economic development and ecological modernization, environmental improvement will occur and thus no institutional environmental arrangement with set cross-boundary environmental goals is necessary. Environmental improvement will come about through economic cooperation and technological innovation. And the Mediterranean region has made great strides in economic cooperation through the Mediterranean Free Trade Zone agreement. Another great partner in trade liberalization has been the EU, which has a special relationship with most Mediterranean Basin states and is one of the most important actors in bringing about economic integration. In addition, it is one of the strongest environmental actors globally.

One contribution by the EU to the Mediterranean EU member states has been the harmonization of legislation through the EU legal framework. Spain, France, Italy, and Greece have benefited from pollution standards and guidelines as well as the Blue Flag scheme for clean beaches. These standards have been negotiated by all EU member states and are also applicable to new members. Often accused of operating at the lowest common denominator, the EU environmental framework is nevertheless one of the strongest environmental codes globally and has without a doubt improved the state of the environment in the northern rim states. Although the EU entertains cooperative networks with most Mediterranean states, this does not extend to imposing EU environmental standards elsewhere. Therefore, the benefit of the EU legislation is limited to its member states in environmental and more widely economic terms. Despite this limited geographic reach, the EU can be seen as one of the most active environmental actors in the region. Given that the EU member states are also the most industrialized states in the Mediterranean Basin and that industrial pollution represents a major share of the overall pollution, an effective EU environmental policy has a substantial impact on the quality of the Mediterranean Sea. Improvements in the state of the Mediterranean as a result of EU policies, however, cannot be attributed to the success of MAP—although they happened during MAP's life span. The EU and MAP II work hand in hand on promoting the principles of sustainable development, and the EU has a strong commitment to free markets.

As such, the last twelve years under the new MAP have revealed a political economy framework that firmly remains an enabling agreement rather than an organization that spawns large-scale interstate cooperation in environmental matters. It has evolved to be part of a much larger structure of institutions and away from being primarily engaged in knowledge production. Thus, it can be seen as a part of a much larger blueprint in terms of the overall development strategy in the Mediterranean region. This firmly takes environmental cooperation as a single issue into the much wider field of trade liberalization and economic development, of which environmental concerns are an integral part rather than a separate issue.

Conclusion

In this chapter I have tried to bring together the historical milestones of MAP along with a conceptualization of these achievements. It is true that the birth of MAP's institutional apparatus was motivated by scarcity in clean seawater. Yet the evaluation of its success depends on how one defines its usefulness or effectiveness. If MAP is seen primarily as an enabling institution in terms of confidence-building measures and a forum for dialogue over time on the issue area of the environment as well as a knowledge-generating institution, then undoubtedly it has been a successful agreement. If it is seen as an institution that is supposed to facilitate cooperation over issues arising from environmental degradation, then it can be described as a partial success. It seems that as a cooperative arrangement, MAP has been quite successful, even though this cooperation has not been put to the test, and takes place at the negotiation table more than in actual joint policies or programs. While there is a joint research agenda, projects are predominantly established at the national level or below, and not across borders—and especially not across borders in conflict areas. Part of the reason for this is that the scientific research generation by MAP indicates that most environmental problems in this semienclosed sea are of a nature that does not necessitate common pollution-abating policies, even if many of them are of a transboundary nature. But cooperation is required for researching the root causes of pollution and pooling resources on knowledge generation. In other words, while joint research was motivated by pollution and the clear degradation of the sea, the initial scientific results suggested that the pollution remained fairly local. To that extent the level of perceived scarcity, and hence urgency to deal with the problem in a coordinated fashion, was effectively reduced. That being said, scarcity and degradation have nonetheless been at the root of continued normative coordination through scientific works, research, and capacity building (environmental concern married with economic development) at the national level. The latter, in particular, is considered important for assuaging the effects of localized pollution. While political, economic, and ideological asymmetries did not play a guiding role in making MAP a normative rather than a policy-driven institution, they certainly contributed to the fact. After all, the economic geography of the Mediterranean region as well as its ideological divisions left the social relations between the member states in danger of being uneven from the start, and while the agreement was successful in not letting these asymmetries dominate the institution, they were still a reality.

In fact, MAP's other mission has evolved into an ideological one, highlighting the compatibility between the environment and economic development. If one evaluates the success of MAP and the Mediterranean Commission on Sustainable Development in terms of how well they integrate environmental concerns into actual policies, and how MAP puts into practice the suggestions of its policy planning programs, which address the socioeconomic level of analysis, then unfortunately this institution has not been successful at all. While MAP does generate research that addresses the roots of environmental problems—namely, nature-society relations—this research is published merely as recommendations and findings, and does not filter through to the policy level at all. Thus, MAP II cannot be described as a policy-oriented institution; it instead remains focused on normative issues as an enabling institution and has transformed itself into a supporting institution for a much wider framework rather than serving as a vital structure in its own right. It acts out an ideological commitment to sustainable development through neoliberal policies, and is committed to distributing these values with the help of other institutions such as the EU and multilateral bodies. This is a particular form of cooperation, but one that is not environmentally driven. The focus on the neoliberal approach to protecting the environment finds its roots in the organization of the global economy, but also in uneven—that is, asymmetrical—relations.

Notes

1. Available at <http://195.97.36.231/dbases/webdocs/BCP/MAPPhaseI_eng.pdf> (accessed November 21, 2009).

2. For more information on MEDPOL, see <http://www.unepmap.org/index.php?module=content2&catid=001017003> (accessed November 21, 2009). For more information on the Blue Plan, see <http://www.planbleu.org/planBleu/missionUk.html> (accessed November 21, 2009). For more information on PAP, see <http://www.pap-thecoastcentre.org/> (accessed November 21, 2009).

3. Officially known as the Protocol for the Prevention of Pollution in the Mediterranean Sea by Dumping from Ships and Aircraft, February 16, 1976, available at <http://195.97.36.231/dbases/webdocs/BCP/ProtocolEmergency76_eng.pdf> (accessed November 21, 2009).

4. Officially known as the Protocol concerning Cooperation in Combating Pollution of the Mediterranean Sea by Oil and Other Harmful Substances in Cases of Emergency, February 16, 1976, available at <http://195.97.36.231/dbases/webdocs/BCP/ProtocolEmergency76_eng.pdf> (accessed November 21, 2009).

5. Officially known as the Protocol for the Protection of the Mediterranean against Pollution from Land-Based Sources, available at <http://195.97.36.231/dbases/webdocs/BCP/ProtocolLBS80_eng.pdf> (accessed November 21, 2009).

6. Officially known as the Protocol concerning Mediterranean Specially Protected Areas, April 3, 1982, available at <http://195.97.36.231/dbases/webdocs/BCP/ProtocolSPA82_eng.pdf> (accessed November 21, 2009).

7. Officially known as the Protocol for the Protection of the Mediterranean Sea against Pollution Resulting from Exploration and Exploitation of the Continental Shelf and the Seabed and Its Subsoil, October 14, 1994, available at <http://195.97.36.231/dbases/webdocs/BCP/ProtocolOffshore94_eng.pdf> (accessed November 21, 2009).

8. Officially known as the Protocol concerning Specially Protected Areas and Biological Diversity in the Mediterranean, June 10, 1995, available at <http://195.97.36.231/dbases/webdocs/BCP/ProtocolSPA9596_eng_p.pdf> (accessed November 21, 2009). Officially known as the Protocol on the Prevention of Pollution of the Mediterranean Sea by Transboundary Movements of Hazardous Wastes and Their Disposal, October 1, 1996, available at <http://195.97.36.231/dbases/webdocs/BCP/ProtocolHazardousWastes96_eng.pdf> (accessed November 21, 2009).

9. Available at <http://195.97.36.231/dbases/webdocs/BCP/MAPPhaseII_eng.pdf> (accessed November 21, 2009).

10. A product of the Fourth Ordinary Meeting of the Contracting Parties to the Convention for the Protection of the Mediterranean Sea against Pollution and Its Related Protocols, September 13, 1985, UNEP/IG/.56/5 (1985), available at <http://195.97.36.231/acrobatfiles/85IG56_5_Eng.pdf> (accessed November 21, 2009).

11. Ibid., reprinted at <http://www.unu.edu/unupress/unupbooks/uu15oe/uu15oe0h.htm> (accessed November 27, 2009).

12. Officially known as the Nicosia Declaration: Charter on the Euro-Mediterranean Cooperation concerning the Environment in the Mediterranean Basin, April 28, 1990. Additional information is available at <http://www.unepmap.org/index.php?module=library&mode=pub&action=view&id=13324> (accessed November 27, 2009).

13. Available at <http://www.pap-thecoastcentre.org/about.php?blob_id=22&lang=en> (accessed November 21, 2009).

14. Unpaid pledge estimate: US$1,540,814. Mediterranean Trust Fund contribution commitments: US$4,406,325.

15. All the revised and amended protocols are available at <http://www.unepmap.org/index.php?module=content2&catid=001001001> (accessed November 21, 2009).

16. Additional information is available at <http://ec.europa.eu/external_relations/euromed/index_en.htm> (accessed November 21, 2009).

References

Bächler, Günther. 1999. "Environmental degradation in the South as a cause of armed conflict." In *Environmental Change and Security: A European Perspective*, ed. Alexander Carius and Kurt Lietzmann, 107–129. Berlin: Springer.

Benoit, Guillaume, and Aline Comeau. 2005. *A Sustainable Future for the Mediterranean: The Blue Plan's Environment and Development Outlook*. London: Earthscan.

Chircop, Aldo. 1992. "The Mediterranean Sea and the Quest for Sustainable Development." *Ocean Development and International Law* 23 (1): 17–30.

Convention for the Protection of the Marine Environment and the Coastal Region of the Mediterranean [known above as MAP II]. 1995. June 10.

Available at <http://195.97.36.231/dbases/webdocs/BCP/bc95_Eng_p.pdf> (accessed November 21, 2009).

Convention for the Protection of the Mediterranean Sea against Pollution [known above as the Barcelona Convention]. 1976. February 16. Available at <http://195.97.36.231/dbases/webdocs/BCP/BC76_Eng.pdf> (accessed November 21, 2009).

Dabelko, Geoffrey, ed. 2001. *Environmental Change and Security Project Report* 7. Washington, DC: Woodrow Wilson International Center for Scholars.

Gillespie, Alexander. 2001. *The Illusion of Progress*. London: Earthscan.

Grenon, Michel, and Michel Batisse, eds. 1989. *Futures for the Mediterranean Basin: The Blue Plan*. Oxford: Oxford University Press.

Haas, Peter. 1989. "Do regimes matter? Epistemic communities and Mediterranean pollution control." *International Organization* 43 (3): 377–403.

Haas, Peter. 1990. *Saving the Mediterranean*. New York: Columbia University Press.

Hoballah, Arab. 2006. "Sustainable development in the Mediterranean region." *Natural Resources Forum* 30 (2): 157–167.

Homer-Dixon, Thomas. 1999. *Environment, Scarcity, and Violence*. Princeton, NJ: Princeton University Press.

Kütting, Gabriela. 2007. "Environment, development, and the global perspective: From critical security to critical globalization." *Nature and Culture* 2 (1): 49–66.

Lempert, Robert, and Gwen Farnsworth. 1994. "The Mediterranean environment: Prospects for cooperation to solve the problems of the 1990s." *Mediterranean Quarterly* 5 (4): 110–124.

Liotta, Peter. 2003. "The uncertain certainty: Environmental stress indicators and the Euro-Mediterranean space." *Mediterranean Quarterly* 14 (2): 21–45.

Massoud, May A., Mark Scrimshaw, and John Lester. 2003. "Qualitative assessment of the effectiveness of the Mediterranean Action Plan: Wastewater management in the Mediterranean region." *Ocean and Coastal Management* 46 (9): 875–899.

Raftopoulos, Evangelos. 1993. *The Barcelona Convention and Protocols: The Mediterranean Action Plan Regime*. London: Simmonds and Hill.

Redclift, Michael. 1987. *Sustainable Development: Exploring the Contradictions*. London: Routledge.

UNEP/IG.18/4.1979. Report of the Executive Director on the Implementation of the Mediterranean Action Plan. Annex III. Athens.

UNEP (OCA)/MED IG.5/16. 1995. Report of the 9th Ordinary Meeting of the Contracting Parties to the Convention for the Protection of the Mediterranean Sea against Pollution and Its Protocols. Annex XIII. Athens.

UNEP (United Nations Environment Programme), MAP (Mediterranean Action Plan), and PAP (Priority Actions Programme). 2001. *Good Practices Guidelines for Integrated Coastal Area Management in the Mediterranean*. Split, Croatia: Priority Actions Programe.

IV

Scarcity and Degradation of Renewable and Nonrenewable Resources

7

Degradation and Cooperation on the High Seas: The Case of International Fisheries Management

J. Samuel Barkin

High seas fisheries were once seen as essentially limitless resources. Fishing in the open ocean, thousands of miles from home, was hard and dangerous work, but the risks did not include stock depletion. Modern fishing technologies, however, have created a condition of scarcity on the high seas. A focused effort by one of the world's many large fishing fleets can deplete a species in a matter of years, and it is often not clear that the species is being overfished until there is a precipitous drop in its population. This puts a particular burden on international cooperation to manage these fisheries, for without cooperation both specific fish species and entire regions of the ocean can be degraded to the point where they neither provide economically viable resources nor sustain viable marine ecosystems.

The onset of the degradation of a fish stock may indeed motivate the relevant international actors to cooperate to manage the fishery. Yet it may likewise motivate none of the actors to cooperate, in which case further degradation is likely, often to the point where the fishery ceases to be economically or even biologically viable. There is nevertheless a third possible outcome, when some states are motivated to manage the fishery effectively but others are not. This situation often leads to escalation, when states begin to treat fishery management as a political issue rather than an issue for scientific or functional management. Escalation may in many cases be a necessary route to get from degradation to cooperation, and as such, is a key focus of this chapter.

The case of international cooperation to manage fisheries resources combines features that are found in many issues of international environmental politics. The fisheries in question are either located on the high seas, a global commons, or are transboundary, and therefore cannot be effectively regulated by one country. They are common resources, meaning that while there is a general incentive to manage the fisheries sustainably, individual fishers have an incentive to fish as much as possible, before the stock is depleted. There is scientific uncertainty about how much fishing a stock or region can support sustainably, and there is political disagreement about who should get to do the fishing that can be supported sustainably. There are tensions between the needs for sustainable management of fisheries, and the economic and cultural demands of fishers and the consumers of fish. And once cooperative agreements are reached, there are problems of cheating, monitoring, and enforcement that plague so many treaties and international institutions (for a general overview of these issues, see DeSombre 2007).

This chapter focuses on those elements of the relationship among scarcity and degradation, conflict and cooperation, that are most specific to international fisheries issues. It does not attempt to address all of the relevant concerns. For example, it does not discuss cheating and enforcement issues, which are indeed important to successful international cooperation to manage the world's fisheries, but that are similar to the equivalent issues in other areas of international environmental cooperation. What this chapter does concentrate on is the interrelationship of the resource and commons aspects with the political economy of international fisheries cooperation. Fisheries represent the only contemporary international environmental issue that involves a resource in a global territorial commons that is being economically exploited to the point of degradation. It is this combination that makes the issue unique.

I will begin by presenting the concepts of degradation, conflict, and cooperation that are at the intellectual core of this volume as a typology of political outcomes in attempts to manage international fisheries. I will then discuss the concept of common pool resources, which brings together both the resource and commons aspects of the issue. In this context, I will also highlight the various asymmetries applicable to the fisheries case, including differing levels of wealth among the parties concerned, the parties' differing shadow of the future with regard to the well-being of the fisheries stock, the degree of a country's dependency on the fish stock, and the directional nature of the fisheries in question. As a domestic counterpoint to the international characteristics of the issue, the following section will explore some of the characteristics of the domestic politics of international fisheries issues. Both the commons and domestic characteristics of these issues feed into the characteristics of international bargaining over the management of fisheries resources, which in turn affects whether depletion will lead to degradation, conflict, or cooperation. I will then present a few brief historical examples to illustrate these dynamics. A final section will note a key limitation on international fisheries cooperation—one that threatens to undermine effective cooperation and promote degradation. This limitation is the tendency, for reasons relating to the domestic politics of fisheries issues, to focus on the process of fishing rather than on the supply of fishers. Without more effort to address the global oversupply of fishers, neither cooperative nor conflictual approaches to international fisheries issues will be able to address the scarcity that underlies resource degradation.

Degradation, Conflict, and Cooperation

Environmental scarcity is a condition in which we want more of a resource than we can use sustainably. Prior to the twentieth century, high seas fisheries were not, in a human sense, scarce, because our technology could not harvest them at an unsustainable rate. Many specific fish stocks close to shore did become scarce since they could be fished out with preindustrial technologies. But the ocean deeps were boundless in their bounty. This remained true until the introduction of industrial technology and particularly the internal combustion engine to fishing.[1] Slowly, both fishers and governments came to realize that factory fishing could indeed degrade those stocks previously thought to be limitless. Responses varied, including enclosure, cooperation, war, and the depletion of some fish stocks to the point where commercial fishing was no longer viable (Montaigne 2007). As fishing technology continues to improve, fish become increasingly scarce, allowing humans to degrade the resource in a matter of years instead of decades. As this relative scarcity increases, questions of degradation, cooperation, and conflict become more pressing.

To this point I have referred both to high seas fisheries and international fisheries management. High seas fisheries, strictly speaking, are only one of the sorts of fisheries that require international cooperation. International fisheries management includes four categories of fisheries. They are pure high seas fisheries, straddling stocks fisheries, migratory stocks fisheries, and inshore fisheries. The last of these categories, which would, for example, include international agreements that set standards for the management of coral reefs, is not covered by this chapter.[2] The other three categories involve fish species that either populate the oceans outside territorial waters and national exclusive economic zones, or cross national maritime boundaries, or both. High seas stocks are those that live entirely in international waters, including some groundfish species like turbot or Patagonian toothfish (which most people know by the name under which it is marketed, Chilean sea bass). Straddling stock fisheries are those in which stocks are to be found both in territorial waters and the high seas, and highly migratory stocks are those in which individual fish move regularly between national waters and the high seas. Different categories generate different international bargaining dynamics, as will be noted in this chapter, where relevant. For instance, migratory stocks generate directional issues, whereas high seas stocks do not (this will be explained below). That being said, most of the argument developed in this chapter applies to all three categories.

All international fisheries resources are scarce, in the sense that all could easily be depleted with modern fishing technology. This is not to say that all are currently under threat—some of these resources are presently being managed sustainably, and others are simply not being fished beyond their maximum sustainable yield. But some species in this latter group have been overfished in the past and may be again in the future (see, for example, the discussion of krill in Joyner 1999). When does scarcity lead to degradation? When the resource is being used beyond the rate at which it can replenish itself (it can also be degraded by habitat loss, which is a real problem for many marine species, but not one that will be addressed in this chapter). By this definition, nonrenewable resources are degraded by any use. But renewable resources are degraded only by use beyond a certain point. It is a characteristic of fisheries that once this point has been reached, degradation can be both quick and precipitous—once overfishing has passed a certain point, the size of a fish stock can collapse from one year to the next (Wilson et al. 1994).

The degradation of international fisheries resources is thus both a biological and an economic phenomenon. It is biological because it reflects the breeding characteristics of the species, and it is economic because it reflects the intensity of the fishing effort. Yet it is also political inasmuch as the degradation of a fish stock, or a marine ecosystem, reflects the failure of political authorities to manage the economics of fishing to match the biology of the fish. As I am using the terms here, therefore, scarcity is a background condition, given a certain level of technology and demand for a resource, while degradation is an outcome—one that reflects the failure of international environmental politics to manage the resource. It is one of three categories of outcomes that I will discuss in this chapter. The other two are also core concepts of this volume: cooperation and conflict.

It is a defining characteristic of the fisheries in question here that they cannot be managed effectively by one state. States realize this, and do not expect that unilateral management will be effective. As such, when confronted by evidence that scarcity is leading to degradation in particular fisheries, states can do one of three things: nothing, negotiation, and escalation. Negotiation refers to attempts to negotiate with other relevant states to come to a mutually agreed on mechanism for managing the fishery, usually through the creation of an international institution of greater or lesser formality, such as a regional fisheries management organization (RFMO).[3] Escalation refers to actions beyond the immediate scope of the management of the particular fishery in question; such actions are designed to encourage other states to negotiate or to change the terms on which they do so.

Escalation can take a number of forms. It can involve positive or negative incentives. Positive incentives are anything that compensates other states for agreeing to a management regime that they would otherwise not agree to, including side payments or concessions in negotiations on different issues (Baldwin 1971). Negative incentives are anything used to threaten consequences to states that fail to agree to a management regime. These incentives range from bad publicity, to exclusion from markets, to trade sanctions, to the use of force.[4] This entire range of incentives, both positive and negative, has been used at various points as a response to scarcity and degradation in international fisheries over the past few decades. Other things being equal, positive incentives are most likely when interest in the fishery is economic rather than cultural, when the target of the incentives has a tradition of effective fisheries cooperation, and when cooperation is most likely to be self-enforcing (all three of these conditions will be defined and discussed in more detail below).

Many of the activities defined here as escalation might be thought of in other contexts as elements of negotiation. For example, positive incentives such as side payments might be considered a standard negotiating tool, as might negative incentives designed to coerce, such as gunboat diplomacy. The term negotiation is used in this chapter in a particularly narrow sense, however, to capture the distinction between attempts by a group of states that are all motivated by an interest in the effective management of a fishery to create an effective means for doing so, and attempts by some states to create effective management regimes against the opposition of other states.

The three policy responses to degradation—doing nothing, negotiation, and escalation—can in turn yield three categories of outcomes: degradation, cooperation, and conflict. The policy responses do not map neatly onto the outcomes. Doing nothing will generally lead to degradation, barring sudden exogenous changes in the marine ecosystem or in fishing efforts. Negotiation may well lead to cooperation, if it is successful. Escalation will likely lead to conflict, if it involves negative incentives. Positive incentives, if successful, may lead directly to cooperation, or if unsuccessful, to degradation. It must also be noted that the three categories of outcomes are not mutually exclusive. Cooperation may well not prevent degradation, if the resultant agreement fails to limit fishing efforts sufficiently.[5] Conflict may continue over time, but may also resolve itself more or less quickly into cooperation, if the negative incentives involved in the escalation succeed in convincing the target party to change its behavior. And there are circumstances under which conflict might even exacerbate degradation.[6]

When are states likely to negotiate as a first choice? When are they likely to do nothing, and when are they likely to escalate? This depends on a number of factors, including both the characteristics of the international fishery and those of the domestic industry and regulatory structure. Looking first at the characteristics of the international fishery, these can be addressed through the analytic lens of common pool resources, and asymmetries in states' time horizons with respect to those resources.

Common Pool Resources and Time Horizons

International fisheries resources, like all environmental issues in international politics, are common pool resources (Barkin and Shambaugh 1999). These are goods that in terms of the characteristics being discussed here are rival but nonexcludable. It is difficult to prevent access to them because until they are harvested, they are not owned by anyone, yet they can be used up. That international fisheries are common pool resources has two key effects, one economic and one political, that create problems for sustainable management. The economic effect is that it creates an incentive for fishers to overfish. If fish were a private good, then fishers who refrained from overfishing today would know that they would be able to benefit from a healthy stock tomorrow. Many domestic fishery management regimes in fact attempt to give fishers effective ownership of their stock for this reason. But because resources outside national jurisdictions cannot be owned by anyone, fishers have no way of knowing who will benefit from their restraint. So there is no economic incentive for restraint. In fact, there is an economic incentive to over-exploit the stock as quickly as possible, so that other fishers do not use it up first (Barkin 2004; Barkin and Shambaugh 1999).

This incentive to overfish is what generates the need for international management. If there were a central authority to manage international fisheries that acted in the long-term interest of the industry as a whole, then it might be able to regulate fishing in a way that ensured the long-term health of fish stock and marine ecosystems. This is what governments are supposed to do for domestic industries, but often fail to accomplish because they either choose not to or are unable to act in the long-term interests of the fishery.[7] But the problems of domestic regulation are compounded by the absence of a central body with the ability to authoritatively regulate. This means that any effective regulation requires cooperation among national governments. While these governments presumably share some interest in the long-term health of marine ecosystems, they also have an interest in promoting their own fishing industries and fleets over those of other countries. It is here that the second and political effect of common pool resources becomes relevant.

Given that national governments are likely, to some extent at least, to prioritize their own national interests, the degree to which any given government is willing to forego a current ability for its nationals to fish a particular stock will be related to the degree to which it is concerned about the short versus the long term. A government that cares more about long-term management (a government that in economic parlance has a low discount rate, or in international relations theory parlance has a long shadow of the future) should be more willing to make short-term concessions, in the form of limiting the amount that its nationals can fish, in exchange for a management regime that will ensure the long-term health of the fishery. A government that cares more about the short-term benefit to be derived from fishing (with a high discount rate, or a short shadow of the future) should be less willing to make short-term concessions for the sake of long-term sustainability. In other words, that international fisheries are common pool resources means both that international cooperation is necessary to manage them, and that to the extent that countries have different shadows of the future, they will have different levels of interest in making the short-term sacrifices necessary to do so.

What would give a country a shorter shadow of the future for international fisheries management? There are a number of factors that can contribute to this, including the country's general level of wealth, the relationship between its fishers and the particular fish stock or marine ecosystem in question, and the sociopolitical context of the fishery in that country.

The Relative Level of Wealth

The first of these factors, the general level of wealth, is relatively straightforward. Other things being equal, richer countries are more likely to be concerned with the long term relative to the short term than are poorer countries. Richer countries are more able to deal with the short-term costs of cooperation and have a greater economic cushion with which to deal with any economic dislocation caused by the creation of a system of sustainable management (Haas, Keohane, and Levy 1993, 404–408). Note that this observation is a general rule; some poorer countries do focus successfully on long-term management, and there are certainly richer countries that are not as concerned about sustainability as they might be. But in aggregate, the observation about the level of wealth and the shadow of the future seems to hold (as is true of individuals as well; see Lawrance 1991).

Substitutability

The second of the factors is the relationship between a national fishing fleet and the particular stock or ecosystem in question. This factor can be summed up with the term substitutability. The extent to which a country is committed to the health of a specific stock is likely to be related to the dependence of its fishers on that stock. The higher the degree of dependence, the longer the shadow of the future. In other words, the more easily the fishers can substitute another fish stock for the stock in question, the less they are likely to be committed to the health of that particular stock in the long term. Conversely, the more fishers know that they cannot substitute for that particular stock, that their long-term well-being is tied to that stock, the greater their interest in making short-term concessions to improve the long-term prospects of the stock (DeSombre and Barkin 2000).

Substitutability is in turn determined by a number of factors related to the type of equipment that the fishers are using and the characteristics of the stock. With respect to equipment, key factors include the size and range of the boats. Larger boats, and boats with longer range, are less likely to be dependent on particular species, and even less likely to be dependent on particular marine ecosystems, than are smaller boats and boats with shorter range. To use examples at both ends of the spectrum, large factory trawlers that are able to stay at sea for months, process their catch onboard, and have global range are likely to look at particular stocks as highly substitutable. If a factory trawler of this type can just as easily fish hake off Namibia as turbot off Greenland, it is unlikely to have a long-term commitment to the health of either of those specific stocks. If one gets fished out, they can move to the other. Conversely, a day boat is severely limited in the distance from its home port that it can travel to fish. If its home waters get fished out, there are no other options with that equipment (DeSombre and Barkin 2000).

It might seem at first that this distinction should not be of that much relevance to international fisheries, because fishing on the high seas is likely to be primarily the preserve of long-range boats. This is true with respect to some fisheries. But international fisheries include straddling and migratory stocks as well as pure high seas fisheries, and straddling and migratory stocks are often caught close to shore. For example, tuna, perhaps the most migratory of species, are nonetheless frequently caught close to shore.[8] Pacific salmon regularly cross the U.S.-Canada border and swim far out into the Pacific. But they are generally caught close to shore, and are in fact often caught in their spawning rivers rather than in the ocean (Barkin and Mosely 2003).

Pacific salmon also provide an example of a way in which the characteristics of a species can affect its substitutability. Salmon are anadromous, meaning that they hatch and spawn in freshwater, but spend most of their lives in saltwater, at sea. Furthermore, they spawn in the same river in which they hatch. This means that they have a specific migration pattern, which has in turn generated a fishery that is designed around the pattern. Since there are no other species on the Pacific coast of North America that display similar behavior and could support a major commercial fishery, salmon fishers are tied to their species much more than fishers for many other species (Barkin and Mosely 2003). In other words, from the perspective of Pacific salmon fishers, there are no species that are substitutable for salmon in the way that, say, hake is often a close substitute for halibut (DeSombre and Barkin 2000).

Sociopolitical Context and Culture

The third of the factors affecting a country's shadow of the future with respect to cooperation in international fisheries management is the sociopolitical context of the fishery in specific countries, and in particular the cultural embeddedness of fishing. Cultural embeddedness is both common and problematic for fisheries regulation. By cultural embeddedness I mean the extent to which local communities define themselves in relation to fishing in general and often to particular species. Examples can be found from the millennium-old tradition of high seas fishing in the Basque and Galician parts of Spain, to the association of Massachusetts with cod and British Columbia with salmon (Kurlansky 1997; Barkin and Mosely 2003). This association seems at first to suggest a long shadow of the future, for if a community defines itself by its fishery, it should have a long-term commitment to that fishery. But fisheries management frequently means fishing less, and this can decimate fishing communities. In response, governments often take the short-term needs of the community into account in setting quotas, and subsidize either the communities or the fishers (see, for instance, Milich 1999). Inasmuch as these subsidies increase the number of fishers and the capacities of the fishery beyond what the stock can support (and often beyond what the market can bear), it generates an incentive for short-term thinking, because a fleet that cannot afford to stay afloat until the next season will care less about long-term management than a fleet that can (see, for example, Barkin and Mansori 2001).

Bargaining and Power

Differing Shadows of the Future

How does this discussion of shadows of the future affect international cooperation and conflict over the management of international fisheries? When all of the actors involved have short shadows of the future, the outcome is likely to be further degradation, until the point is reached that the fishery is no longer economically viable. Since the actors in this circumstance are more interested in getting what they can from a fishery in the short term than they are in managing it sustainably in the long term, they have no incentive to negotiate. Conversely, when all actors have long shadows of the future, negotiation leading to cooperation is likely. But this circumstance is not common in high seas fisheries. Since such fisheries are open access, a lucrative fishery is likely to attract opportunistic actors with short shadows of the future. This leads to a situation in which different actors have different shadows of the future, and it is such situations that often result in escalation.

A short shadow of the future gives an actor bargaining power in negotiations about common pool resources with respect to an actor with a relatively longer shadow of the future. In the context of fisheries cooperation, this means that a country that cares more about its short-term quotas will have greater bargaining leverage than a country that cares more about long-term sustainabililty. Because the former country does not care as much (relatively) about the state of the fishery in the future, depleting the fishery is less costly to it than to a country that cares more about the long term. Meanwhile, as the countries are negotiating, the stock is being fished and frequently depleted. The longer the negotiations last, the more the stock is depleted during this process, and the less quota there is to share once an agreement is reached (Barkin 2004).

All of this means that delaying, or holding out in, negotiations is more costly to the country that is focused on sustainability than to the country more focused on short-term gain. This in turn means that the short-term-focused country can credibly threaten to delay negotiations in a way that is more costly to the other country, giving the country more concerned about sustainability an incentive to settle quickly on unfavorable terms rather than drag out negotiations to win better terms. In other words, the country more concerned about sustainability is likely to be faced with cooperation on the other country's terms or depletion of the fishery. If the other country's best offer is one in which its own fishers alone fish at a rate greater than what is sustainable, the country in question may well be faced with a choice between an agreement that both freezes out its own fishers and fails to manage the fishery sustainably, or no agreement, in which case the fishery is depleted even more quickly (Barkin 2004).

It is at this point that the potential for escalation is at its highest. The country that is more concerned with sustainability, faced with unfavorable bargaining dynamics, may well decide that its power disadvantage within the context of the politics of the specific fishery needs to be counterbalanced by power advantages elsewhere, such as market, financial, diplomatic, or military power. The source of escalation, of any form, in international fisheries affairs is almost always an attempt to overcome a bargaining disadvantage, often generated by different shadows of the future in a common pool resource setting. Other things being equal, escalation is more likely the greater the difference in shadows of the future, the greater the outside power resources that can be brought to bear by the country with the greater concern for long-term sustainability, and the greater the level of concern by that country about the stock or ecosystem in question. This concern, in turn, can be generated by either the economic or cultural importance of the fishery, or simply the extent to which the issue catches the popular imagination.

Of the various kinds of escalation available, positive incentives, meaning side payments or concessions in other negotiations, are likely to be most effective when the core interests of the negotiating parties are primarily economic rather than cultural (side payments can also be used as a face-saving mechanism for countries with less negotiating power). Of the various potential negative incentives, the choice of a particular vehicle for escalation depends largely on the power resources available to the country that chooses to escalate. For example, the United States and the European Union (EU) are more likely to use access to their domestic markets as a bargaining tool than are other countries, because they represent the world's two biggest markets for most goods, and therefore the use of access to their markets as a power resource in bargaining is more likely to be effective than it is for other countries. Attempts to use embarrassment of recalcitrant countries as a power resource in negotiations is only likely to work when there is a target audience that is likely to respond that has some impact on political decision making in the target state. And the use of direct force is generally most likely to be effective when it is used close to home waters (note that the use of force here usually means law enforcement rather than military engagement, although the latter has been known to happen).

The source of decisions to escalate is a power disadvantage in negotiations, and as noted above is often generated by differences in time horizons, which in turn are frequently related to differences in general levels of wealth, in stock substitutability, or in the cultural embeddedness of the fishery. But there are other sources of power differentials in negotiations over the management of international fisheries as well. Two of these sources are directionality and federalism.

Directionality

Directionality refers to the extent to which the fishery in question has an upstream/downstream dynamic rather than a pure commons dynamic. In a pure commons dynamic, all countries that use a resource have equal access to it. But if the resource is directional, like a river, then the upstream country has access to it before the downstream country, meaning that the upstream country can simply use up the resource if it chooses before the downstream country has any access to it (DeSombre 2007, 22). This gives the upstream country additional bargaining power, because if it holds out in negotiations, it continues to have access to the resource and can block the access of the downstream country. Pure high seas fisheries have a pure commons dynamic, but many straddling and migratory stocks cross national borders in a way that generates an upstream/downstream characteristic. The Pacific salmon is a good example of this, as will be discussed below.

Political Structures

As well as the characteristics of fish species, the characteristics of political structures can also affect bargaining power. One example of this can be found in federal political structures, in which some or all of the authority to regulate fisheries is vested somewhere other than the federal government. A unitary central government can legally (if not necessarily politically) agree to any sort of international fisheries deal it wants to. But a federal government that does not have authority over fishery resources within a certain distance from shore is constrained in its international negotiations by what the lower level of government will agree to. Perhaps counterintuitively, this situation gives greater bargaining power to the more constrained central government, because that government is less able to make concessions, putting more of the onus for reaching a cooperative agreement on the government that is more able to make concessions (Putnam 1988).

It should be noted here that in the case of negotiations on international fisheries management, it makes some sense to look at the EU as a federal government, and at its member governments as the equivalent of state-level governments. In some ways, the member governments of the EU have similar levels of control over their marine resources as state governments in the United States.[9] It should also be noted that as is the case with directionality, this source of power is only relevant with respect to straddling and migratory stocks, particularly those that come quite close to shore at some point. Negotiations over pure high seas fisheries are always within the purview of national governments, except in the case of the EU.

Negotiation and Escalation, Cooperation and Conflict

Escalation, then, is the response of the country with less bargaining power in negotiations over the management of international fisheries resources when that country is unwilling to accept the alternatives of either continued degradation or cooperation on distributional terms that it cannot accept. A country in this situation will not necessarily escalate. Whether it chooses to do so or not depends on how much it cares about the health of the fishery in question, and whether it has access to alternative sources of power that are likely to be effective. It can also depend on factors such as whether or not domestic lobbying groups or particular politicians choose to pursue the issue. In short, the bargaining power story is one of necessary, but not sufficient conditions.

Escalation as a tool is designed to encourage (or force) international cooperation that prevents or alleviates the degradation of the resource. It will not necessarily succeed in either creating cooperation or forestalling degradation. Uses of power sometimes fail, and cooperation will not necessarily result from escalation. And of course, cooperation itself does not necessarily forestall degradation. But it remains the case that escalation, even through the use of negative incentives that intentionally create a situation of international conflict, is often the best or even only tool available to countries that want to prevent the degradation of international fisheries through cooperative international management.

Three brief examples may help to illuminate some of the dynamics discussed in this chapter. Each case captures the role of particular asymmetries as sources of bargaining power, and together they touch on many of them. The examples include the turbot fishery in the northwest Atlantic in the mid-1990s, the Pacific salmon fishery in the late 1990s, and the Patagonian toothfish fishery in the early years of the twenty-first century.

The Turbot War

Turbot, also known as Greenland halibut, are a groundfish species that straddle the boundary of Canada's exclusive economic zone off the coast of Newfoundland and Labrador.[10] The species had not been fished extensively until the depletion and collapse of the regional cod fishery in the early 1990s. Those parts of the turbot stock in the Canadian exclusive economic zone were regulated unilaterally by Canada (turbot are a straddling stock, but are not migratory), and those in international waters were subject to regulation by an already-existing RFMO, the Northwest Atlantic Fisheries Organization (NAFO).[11] Because the stock had only recently become subject to a major fishing effort, NAFO's first attempt at a quota for it was in 1995. NAFO set a total quota based on a calculation of the maximum sustainable yield, and divided up that quota based on traditional national fishing efforts.

As is the case with most RFMOs, NAFO's rules allowed individual members to object to specific quotas and thereby not be bound by them. The two countries most interested in the turbot quota were Canada, which saw the international turbot fishery as an extension of its traditional Atlantic fishery, and Spain, which had taken the largest catch of turbot in those waters during the previous two years (Spain is represented in NAFO by the EU rather than being a member in its own right). In dividing up the total quota, NAFO gave the largest national quota to Canada, favoring its argument of tradition over Spain's argument of recent practice. The EU objected to its quota and unilaterally announced that it would limit its turbot take to an amount that was below the overall quota, but that in combination with other national quotas would likely result in a serious depletion of the stock. In doing so, the EU was acting within its treaty commitments and international law—RFMOs generally either allow objections or require consensus decision making to encourage participation that is as wide as possible.

Both Canada and Spain are wealthy countries, although the fishing industry in each is concentrated in poorer parts of the country. But turbot were far more substitutable for Spain's fishery than for Canada's. The Canadian fishery was a near-shore one, designed to fish within Canada's exclusive economic zone and able to fish only a little farther out than its border. The fishery within the zone had by then become severely depleted, leaving the fishery with few other options. Spain's fleet, conversely, was deep sea with a global range. As such, Spain's shadow of the future in this case was shorter than Canada's, and Spain was happier with the status quo of unsustainable management than was Canada. Furthermore, the issue was highly politicized, becoming front-page news in both countries.

The Canadian government chose to escalate rather than accede to the higher, unilateral quota that the EU had declared on behalf of Spain. It did so through the use of both force and diplomacy. It proposed a moratorium on fishing turbot until the quota issue was resolved, and announced that it was willing to enforce the moratorium. It then impounded a Spanish trawler that was fishing in international waters (but that was breaking a number of NAFO rules, related to issues such as net size and bycatch). This succeeded in scaring most European trawlers away, at least for the short term. Meanwhile, the Canadian government undertook a diplomatic offensive to counteract the fact that its coast guard had just impounded a ship on the high seas. This was aimed in part at a general international audience, but also in part at other countries in the EU; since it is the EU rather than its member national governments that negotiate fishery quotas, undermining Spain diplomatically with other EU governments had the effect of undermining Spain's negotiating position.

In the end, an agreement was reached that allowed Canada and the EU equal portions of the total quota. It also improved the monitoring system, and tightened rules on equipment and fishing practices designed to improve conservation. Spain, although it accepted the agreement only grudgingly, ended up with a larger share of the quota than it had originally been given, although a smaller share than it had claimed for itself. Canada achieved better rules for sustainable management as well as the ability to fish more than it could have sustainably had it not escalated.

Pacific Salmon

The second example comes from Canada's other coast, and a fisheries dispute with the United States—this one over Pacific salmon.[12] This fishery is of economic importance to the province of British Columbia, and the states of Alaska, Washington, and Oregon, but perhaps more crucially it is also of cultural significance, which magnified the political salience of the dispute. Pacific salmon have a life cycle that makes the fishery different from most others. They hatch in rivers, spend most of their lives at sea, and return to spawn in their home rivers. When at sea, they swim clockwise in a great circle, past Alaska, then British Columbia, then the states of the Pacific Northwest, then out to sea, and eventually back to Alaska. Despite the fact that they spend much of their lives far out to sea, they are usually fished quite close to shore and often in their home rivers as they return to spawn. This gives the fishery an unusual directional characteristic, because both countries are both upstream and downstream of each other.

This convoluted directionality is complicated by the political structure of fisheries regulation in the two countries. Both countries are federal in structure. But whereas in Canada the central government has the authority to regulate fisheries in national waters, in the United States the state-level governments have the authority to regulate fishery resources in territorial waters. This means that negotiations over the management of the Pacific salmon fishery require not only the cooperation of the U.S. government but also the governments of Alaska, Washington, and Oregon. Furthermore, Alaska is upstream of Canada, whereas Washington and Oregon are downstream. All the relevant governments have a long history of cooperating (often unsuccessfully) in the management of the fishery, and Canada and the United States signed an agreement (Pacific Salmon Treaty 1985) that among other things, committed both governments to the principle that salmon quotas were to be divided based on the number of fish that spawned in each country's rivers. In other words, fish belonged to the country in which they were born.

This decision rule seems relatively straightforward, but by 1992 the two countries stopped agreeing on specific quotas. By 1997, the degradation of the fishery was sufficiently clear that the two governments resumed negotiations. Both governments agreed that the salmon harvest should be cut back significantly. The Canadian government argued that this total quota should be divided according to the terms of the existing treaty, but the U.S. federal government did not have authority from the relevant state governments to agree to those terms. This in effect empowered the U.S. government in the negotiations because it could credibly maintain that it did not have the ability to be flexible, whereas the Canadian government did.

The Canadian government escalated the dispute by using legal means to harass U.S. fishers, particularly those in transit between Washington and Alaska through Canadian waters (an action specifically targeted at those fishers downstream of Canada, over whom Canada had more leverage than over Alaska fishers). Various other groups within Canada took action to pressure the United States as well, including a group of fishers in Prince Rupert that blockaded a U.S. ferry in port in protest. Ultimately, however, the Canadian government was limited in the ways in which it could feasibly escalate by its paucity of effective sources of power relative to the United States. The 1997 season ended before an agreement was reached, and negotiations continued. Canada also began negotiating directly with the relevant state-level governments.

By 1999, all of the relevant governments had reached an agreement on the sustainable management of the Pacific salmon fishery. This agreement saw the United States reduce its fishing effort, but Canada reduced its effort by proportionally more. Canada likewise conceded on a key issue: the principle that each country owned the fish spawned in its waters. While the United States retained sole rights to its own fish, Canada retained the rights only to a majority of the fish spawned in its waters. The United States undertook some minor commitments to share in the cost of salmon conservation measures in Canada, but these side payments were largely symbolic, to allow the Canadian government to put a more favorable spin on what was essentially a concession in the core issue of the dispute. Escalation in this case failed to counteract issue-specific bargaining power, yet because both sides began with relatively long shadows of the future, a sustainable management regime was agreed to nonetheless.

Patagonian Toothfish

The final example concerns Patagonian toothfish, marketed under the name Chilean sea bass, as noted earlier.[13] This species was not subject to large-scale exploitation until the 1990s. But since the stock is found in a remote part of the South Pacific, in Antarctic waters, it is only fished by large factory trawlers. As such, once large-scale exploitation began, depletion began soon after. This example is not a direct diplomatic dispute between particular countries. Rather, it is an attempt to combat what is often referred to as illegal, unreported, and unregulated (IUU) fishing. Because international fisheries are common pool resources, fishing outside a regulatory structure is not only a nuisance but a fundamental threat to successful management as well. To the degree that nonparticipants in regulation can themselves cause the depletion of a resource, sustainable management requires that the problem of IUU fishing be addressed.

This story involves two categories of players. The first of these is the major fishing powers, cooperating through the relevant RFMO, the Convention for the Conservation of Antarctic Marine Living Resources (CCAMLR 1980). This category includes countries that also happen to be the primary markets for Chilean sea bass, led by the United States and the EU. The second category is IUU fishers. This includes ships flying the flags of CCAMLR members who break the rules, catch more than they are allowed to, or fail to report what they catch. But it also includes ships flying the flags of nonmembers. These nonmembers are often what is called flags of convenience—countries that allow shipowners who have no direct relationship with the country to register in that country in exchange for certain fees and taxes (thereby to fly its flag, hence the term flags of convenience), but that tend to regulate the ships to lower standards as other flags. These flags of convenience often do not participate in many RFMOs, and long-range fishers register with them specifically to avoid being affected by national quotas.

CCAMLR members agreed on quotas for the stock shortly after large-scale exploitation started. This is an example of actors with shadows of the future that are both similar and relatively long, negotiating and cooperating as a first response to the threat of degradation. Yet the management efforts were undermined by IUU fishing, which in some years was estimated to take in a larger catch than did legal fishing. The problem of illegal and unreported fishing was susceptible to monitoring mechanisms and greater efforts by national governments to hold their fishers to their own rules. Unregulated fishing presents a different sort of problem, however, because all countries have rights of access to high seas fisheries, even countries that are flags of convenience. But these countries have short shadows of the future, for a variety of reasons. They are generally poor countries, and the ships that fly their flags have no long-term commitment to them. Nor are these ships part of the national culture in these countries. Therefore, other things being equal, they have little incentive to participate in international management. And as long as they flag fishing vessels, but do not participate in management, unregulated fishing can undermine sustainable management.

There are different ways to deal with this problem. All involve escalation of some sort, although the escalation is often intended to discourage fishing directly, rather than to encourage negotiation. In the case of the turbot fishery discussed above, for instance, the Canadian government dealt with flag-of-convenience fishing with force: it gave itself the right to impound vessels fishing in international waters outside its Atlantic exclusive economic zone that were flying flags of nonmember countries of NAFO. But this strategy is less feasible in the far more remote Antarctic seas region. So the members of CCAMLR decided to use market force instead. They created what is called the Catch Documentation Scheme, or CDS.[14] This scheme allows regulated fishers to document that their fish were caught within the rules. The participant countries then agreed to prohibit the landing or importation of any Patagonian toothfish the provenance of which could not be documented through the CDS. Since these countries constituted most of the market for the fish, the price of IUU fish fell relative to that of legal, regulated fish. This in turn gave an incentive to fishers either to flag with a member of CCAMLR or to fish elsewhere.

In other words, the CDS created a club. Since club goods are in a way the opposite of common pool resources, attaching the latter to the former can be an effective management tool. The incentive to participate in the club helps to offset the incentive to nonparticipation in the management of the common pool resource.[15] This is another way in which escalation can be used to encourage countries to participate in sustainable management regimes when they would otherwise not do so.

Conclusion

These three examples illustrate the usefulness of concepts such as common pool resources and substitutability in understanding the effects of asymmetries in wealth, resource alternatives, and governmental structures on bargaining power in international fisheries politics. They also highlight the importance of cultural as well as economic factors in generating patterns of conflict and cooperation in the management of international fisheries. They also illustrate that conflict is best seen not as an outcome in itself but frequently as a part of a process leading to cooperation. Nor is cooperation best seen as the solution to resource degradation. It may often be a necessary step in the sustainable management of fishery resources, but that does not make it sufficient. In fact, cooperation that is not preceded by escalation and conflict can be less useful in addressing degradation than cooperation that results from conflict. This was true to at least some extent in all three of these cases. In the turbot case, an existing cooperative arrangement was not sufficient to prevent degradation, even though the participants in this arrangement were acting within the letter of the rules. In the salmon case, an existing agreement was simply being ignored until the impetus of escalation led to a renegotiation and ultimately a more effective agreement. In the Patagonian toothfish case, negotiation and cooperation took place among those actors with similar shadows of the future, yet escalation was necessary to force that outcome on another set of actors who otherwise showed no interest in accepting it.

This chapter has largely focused on the role of cooperative international institutions and agreements in the management of fisheries through rules limiting what fishers are allowed to do, such as quotas, equipment and season regulations, and so forth. In doing so, it has mirrored the large majority of the literature on fisheries in the discipline of international environmental politics. This approach has its limits, however, in that it introduces what might be called a balloon problem. A balloon, if squeezed in one place, simply bulges in another, and continues to maintain the same volume. Similarly, a given number of fishers may well fish a given amount. Efforts to restrict their catch of specific species, or in specific places, while they may have the effect of preventing the degradation of that stock or in that region, may also result in displacing rather than reducing their total fishing effort. To the extent that the overall fishing effort confronting international fisheries is unsustainable, more attention needs to be placed on cooperation to reduce the number of fishers rather than focusing simply on the amount they fish. There are some contemporary attempts to deal with this issue cooperatively, such as negotiations to reduce subsidies to fisheries under the auspices of the World Trade Organization (see, for example, Campling, Havice, and Ram-Bidesi 2007, part 2). But to this point no countries have been willing to escalate to reduce numbers of fishers in a way that mirrors the extent to which they have been willing to escalate to reduce the amount fished of particular stocks.

Ultimately, then, whether dealing with the amount being fished or the number of fishers, escalation and conflict are often the only ways to impact degradation. International conflict is a sign, it is true, that scarcity in international fisheries stocks is a real problem, and has led or is leading to degradation. But it is also a sign that at least some countries care enough about the sustainable management of those fisheries to commit real power resources to encourage international cooperation.

Notes

1. For an engaging description of the evolution of fishing technology and its effect on one particular fishery, see Kurlansky 1997.

2. There is at present no authoritative international cooperative mechanism for the management of coral reefs, although there is the International Coral Reef Initiative with both governmental and nongovernmental membership.

3. For a concise review of the institutional structure of international fisheries management, including RFMOs, see DeSombre 2006, 84–93.

4. On the distinction between negative and positive incentives in international relations, see Levy 1997.

5. To the point where the acronym of the International Commission for the Conservation of Atlantic Tuna (ICCAT), one of the longest-established RFMOs, is sometimes referred to by participants in the international fisheries management community as the International Commission to Catch All the Tuna.

6. If it does not lead to cooperation, and if it has the side effect of shortening actors' shadows of the future, as will be discussed below.

7. For example, when Canada acquired management authority over the bulk of the Grand Banks cod fishery after declaring a two-hundred-mile exclusive economic zone in 1977, it responded by subsidizing an expansion of its fishing fleet and processing capabilities, rather than instituting a sustainable management regime. See Milich 1999.

8. This does not affect the management of the fishery, however, as RFMO agreements stipulate that national quotas apply to stocks whether they are caught in national or international waters.

9. In both cases, the lower level of government has substantial direct control over fisheries management within territorial waters, while control over management in the exclusive economic zones vests at the higher level of government.

10. Material for this case is drawn from DeSombre and Barkin 2002; Barkin and DeSombre 2000.

11. NAFO was established under the Convention on Future Multilateral Cooperation in the Northwest Atlantic Fisheries (1978).

12. Material for this case is drawn from Barkin 2006.

13. Material for this case is drawn from DeSombre 2005.

14. See <http://www.ccamlr.org/pu/e/cds/intro.htm> (accessed November 18, 2009).

15. Club goods are excludable but not rival. Clubs can choose to admit or exclude potential new members. In addition, the more members in the club, the more effective the club can be.

References

Baldwin, David. 1971. "The power of positive sanctions." *World Politics* 24 (1): 19–38.

Barkin, Samuel. 2004. "Time horizons and multilateral enforcement in international cooperation." *International Studies Quarterly* 48 (2): 363–382.

Barkin, Samuel. 2006. "The Pacific salmon dispute and Canada-US environmental relations." In *Bilateral Ecopolitics: Continuity and Change in Canadian-American Environmental Relations*, ed. Philippe Le Prestre and Peter Stoett, 197–210. London: Ashgate.

Barkin, Samuel, and Elizabeth DeSombre. 2000. *The Turbot War: Canada, Spain, and the Conflict over the North Atlantic Fishery.* PEW Case Studies in International Affairs 226. Washington, DC: Institute for the Study of Diplomacy.

Barkin, Samuel, and Kashif Mansori. 2001. "Backwards boycotts: Demand management and fishery conservation." *Global Environmental Politics* 1 (2): 30–41.

Barkin, Samuel, and Cassandra Mosely. 2003. "Sustainable development, political institutions, and scales: The management of Pacific salmon." In *Achieving Sustainable Development: The Challenge of Governance across Social Scales*, ed. Hans Bressers and Walter A. Rosenbaum, 91–108. Westport, CT: Praeger.

Barkin, Samuel, and George Shambaugh, eds. 1999. *Anarchy and the Environment: The International Relations of Common Pool Resources*. Albany: State University of New York Press.

Campling, Liam, Elizabeth Havice, and Vina Ram-Bidesi. 2007. *Pacific Island Countries, the Global Tuna Industry, and the International Trade Regime—A Guidebook*. Honaira, Solomon Islands: Forum Fisheries Agency.

CCAMLR (Convention for the Conservation of Antarctic Marine Living Resources). 1980. May 20. Available at <http://www.ccamlr.org/pu/e/e_pubs/bd/pt1.pdf> (accessed November 18, 2009).

Convention on Future Multilateral Cooperation in the Northwest Atlantic Fisheries. 1978. October 24. Available at <http://www.nafo.int/about/overview/governance/convention/convention.pdf> (accessed November 18, 2009).

DeSombre, Elizabeth. 2005. "Fishing under flags of convenience: Using market power to increase participation in international regulation." *Global Environmental Politics* 5 (4): 73–94.

DeSombre, Elizabeth. 2006. *Global Environmental Institutions*. London: Routledge.

DeSombre, Elizabeth. 2007. *The Global Environment and World Politics*. 2nd ed. London: Continuum.

DeSombre, Elizabeth, and Samuel Barkin. 2000. "Unilateralism and multilateralism in international fisheries management." *Global Governance* 6 (3): 339–360.

DeSombre, Elizabeth, and Samuel Barkin. 2002. "Turbot and tempers in the North Atlantic." In *Conserving the Peace: Resources, Livelihoods, and Security*, ed. Richard Matthew, Mark Halle, and Jason Switzer, 325–360. Winnipeg: International Institute for Sustainable Development.

Haas, Peter, Robert Keohane, and Marc Levy, eds. 1993. *Institutions for the Earth: Sources of Effective International Environmental Protection*. Cambridge, MA: MIT Press.

Joyner, Christopher. 1999. *Governing the Frozen Commons: The Antarctic Regime and Environmental Protection*. Columbia: University of South Carolina Press.

Kurlansky, Mark. 1997. *Cod: A Biography of the Fish That Changed the World*. London: Penguin.

Lawrance, Emily. 1991. "Poverty and the rate of time preference: Evidence from panel data." *Journal of Political Economy* 99 (1): 54–77.

Levy, Jack. 1997. "Prospect theory, rational choice, and international relations." *International Studies Quarterly* 41 (1): 87–112.

Milich, Lenard. 1999. "Resource management versus sustainable livelihoods: The collapse of the Newfoundland cod fishery." *Society and Natural Resources* 12 (7): 625–642.

Montaigne, Fen. 2007. "Still waters: The global fish crisis." *National Geographic* (April). <http://ngm.nationalgeographic.com/2007/04/global-fisheries-crisis/montaigne-text> (accessed December 13, 2007).

Pacific Salmon Treaty. 1985. March 17. Available at <http://www.psc.org/pubs/Treaty.pdf> (accessed November 19, 2009).

Putnam, Robert. 1988. "Diplomacy and domestic politics: The logic of two-level games." *International Organization* 42 (3): 427–460.

Wilson, James A., James M. Acheson, Mark Metcalfe, and Peter Kleban. 1994. "Chaos, complexity, and community management of fisheries." *Marine Policy* 18 (4): 291–305.

8

Conflict and Cooperation along International Rivers: Scarcity, Bargaining Strategies, and Negotiation

Shlomi Dinar

The consequences for humanity are grave. Water scarcity threatens economic and social gains and is a potent fuel for wars and conflict.
—UN Secretary General Ban-Ki Moon, Asia Pacific Water Summit, 2007[1]

Predictions of water wars, such as the one above, continue to resonate today, largely in the popular media.[2] On a whole, scholars have distanced themselves from extreme predictions of water wars, yet ultimately have found themselves hypothesizing similar violent scenarios (Homer-Dixon 1999; Klare 2001).[3]

While it is true that water disputes have taken a military turn on at least seventeen occasions during the period 1900–2001, the last all-out war over water took place forty-five hundred years ago—between the city-states of Lagash and Umma (Hensel, Mitchell, and Sowers 2006, 407; Wolf and Hamner 2000, 66).[4] In comparison, thousands of water agreements have been concluded, with the oldest dating back to 3100 BC.[5] Consequently, as Aaron Wolf and Jesse Hamner (2000, 66) have noted, "the more valuable lesson of international water is as a resource whose characteristics tend to induce cooperation, and incite violence only in the exception."[6]

While violent conflicts over transboundary water may be rare, political disputes and conflicts of interest over shared freshwater are not (Gleditsch 1998, 387). In fact, international disputes over shared rivers—say, due to water allocation or pollution problems—take place on a global scale and are not limited to one region. The environment and security as well as hydropolitics literature has thus turned to explaining why disputes have taken place over these issues, and how cooperation has often succeeded or failed in these contexts.[7]

Among the basic variables often associated with analyzing conflict and cooperation over freshwater is scarcity. Broadly defining scarcity to include issues other than the common topic of water quantity, this chapter first considers how scarcity may lead to conflict and, more important given the theme of this book, cooperation. The section suggests that the relationship between scarcity and cooperation actually follows an inverted U-shaped curve rather than a linear relationship as generally hypothesized. Clearly, other factors are critical for explaining conflict and cooperation over transboundary rivers, and the chapter explores some overarching variables including geographic discrepancies between the river riparians, power differentials among the parties, and the nature of domestic politics among the states. While several other intervening variables may help to explain cooperation along international rivers (i.e., third-party mediators), this chapter contends that since the asymmetries among river riparians (geographic, military/economic, etc.) may often impede cooperation, offsetting such asymmetries is crucial. To that extent, different bargaining strategies are argued to be at the core of understanding how cooperation along international rivers is facilitated. Various examples of interstate regimes in the form of international water treaties are provided.[8]

Scarcity, Conflict, and Cooperation

According to Arun Elhance (1999, 3), "Hydro-politics is the systematic study of conflict and cooperation between states over water resources that transcend international borders." Indeed, the international and transborder characteristics of shared water bodies make them a compelling case for the analysis of conflict and cooperation. River riparians are physically interdependent, because water bodies respect no political borders. The hydrology of an international river basin links all the riparian states, requiring them to share a complex network of environmental, economic, political, and security interdependencies. Therein lies the potential for interstate conflict as well as opportunities for cooperation (Elhance 1999, 13).

While the concept of scarcity in the field of hydropolitics is regularly associated with water allocation or quantity, countries may suffer from other aspects of water-related scarcity including energy, flood protection (flow regulation), or pollution control, and therefore may be likewise inclined to dispute over or manage an international river for hydropower, flood control, or environmental sustainability purposes.

Scarcity and Conflict

Per the conflict side of the hydropolitical coin, scholars have generally argued that in arid regions, where water allocation issues are most pressing, scarcity may be exemplified in the periodic shortages that a nation may experience, which in turn may be intensified by the conflicting uses to which its neighbors have put the river. Consequently, conflicts may easily arise when users (individuals or states) are competing for a limited resource to supply the domestic, industrial, and agricultural sectors (Falkenmark 1992, 279–280, 292).[9] Scarcity compounded by the complex interdependence ascribed to river riparians also places parties in a precarious and potentially volatile situation. In particular, the interdependence elicited by sharing an international river not only highlights the sensitivities between countries but their reciprocal vulnerabilities too. This tends to make cooperation difficult and tensions more likely as states attempt to reduce their dependence on other countries (Waltz 1979, 106, 154–155). The military exchanges that took place between Israel and Syria in the mid-1960s over the parties' respective unilateral water projects to withdraw water from the Jordan River Basin along with the near military showdown between Syria and Iraq in the mid-1970s over the low flows of the Euphrates River are a testament to such scenarios.

Countries may similarly be guided by relative gains concerns and thus shun cooperation, even if cooperation may bestow benefits on all parties. In essence, riparians fear that discrepancies in otherwise mutually desirable gains will overwhelmingly favor treaty partners. In the area of pollution control, for example, the cost of water contamination is often borne by the downstream riparians, abetting a further lack of bilateral cooperation (Kratz 1996, 26). In the area of water quantity and allocation, the failure of the Johnston Plan in 1955 to encourage a basinwide agreement for the Jordan River Basin is a telling example. Both Israel and the neighboring Arab states were concerned that any concessions, suggested by President Dwight Eisenhower's envoy Eric Johnston, would provide substantial benefits to the other, and compromise their own security and capabilities (Lowi 1993, 192–193; Johnston Negotiations 1955).

Several empirical studies have also demonstrated the association between water availability and armed conflict. Hans Petter Wollebæk Toset, Nils Petter Gleditsch, and Håvard Hegre (2000, 992–993), for example, find that while water scarcity is not necessarily the only, or main, issue in explaining armed conflict, the "low availability of water in both countries of the dyad is significantly related to disputes." While Gleditsch and other colleagues (2006, 376) find some ambiguous results for the relationship between water scarcity and interstate conflict, their findings suggest that countries experiencing low average rainfall have a higher risk of interstate conflict. It is important to note that these companion studies consider armed conflict in general, which may or may not be associated with a water dispute. Specifically, focusing on competing claims over cross-border rivers, Paul Hensel, Sara McLaughlin Mitchell, and Thomas Sowers (2006, 390) also conclude that militarized disputes are more likely to take place in more water-scarce regions. The authors contend that resource-poor areas create environments whereby the creation of institutions to manage conflict will be lacking and/or ineffective (Hensel, Mitchell, and Sowers 2006, 385, 388, 408–409).

Scarcity and Cooperation

For the same reasons that scarcity may initiate interstate conflict, it can likewise initiate cooperation (Dokken 1997). Water is necessary for all aspects of national development, and thus cooperation over a shared river can contribute to the welfare of a society and allow states to cope with scarcity. In short, cooperation between countries sharing the same basin will become increasingly critical as water becomes scarcer (Rosegrant 2001, 150).

Overall, such a common interest as alleviating scarcity eschews relative gains maximization (Stein 1982, 318). The plain desire to reach a mutually satisfying outcome in dealing with scarcity provides the initial incentive to form an international regime (Keohane 1982). Consequently, when unilateral efforts are considered inefficient and less beneficial, cooperation and coordination are more likely (Elhance 1999, 50–51; Barrett 2003, xiii). Interdependence therefore does not solely promote a relationship of vulnerability between river riparians, as implied above, but rather a type of relationship where neither riparian may act without some type of coordination with the other party (Burton 1972). For example, if a downstream state demands additional supplies of electricity and wishes to exploit the river's hydropower potential, it may require the upstream state to regulate the river's flow and build the needed dams. Hydropower plants must also be constructed and are frequently located further downstream. With a potential energy source available, the upstream state will look to the downstream state to transfer some of the hydropower produced for its own needs. Since the utilization of the river benefits each party, cooperation is considered a key avenue for such gain.

The Indus River Agreement (1960) constitutes an instructive case of how scarcity may lead to regime formation. This is a particularly important case since by some accounts India and Pakistan should have gone to war over the Indus waters. According to Undala Alam (2002, 342), all the ingredients were present: two enemies engaged in a wider conflict, with one riparian especially dependent on the river, water scarcity, and poverty preventing the construction of infrastructure to offset the scarcity. In this case, where relative gains concerns would seem to be particularly conducive to continued conflict, Alam claims that "both riparians needed water urgently to maintain existing works and tap the irrigation potential in the Indus basin to develop socio-economically." Furthermore, "Pakistan felt that India's upstream developments on the Sutlej River would damage its existing uses and, therefore, threaten its very livelihood." In turn, "India planned to develop its irrigation potential to offset poverty in the country." And so "by signing the Indus Waters Treaty, both countries were able to safeguard their long-term water supplies" (Alam 2002, 347).[10] In a testament to the significance of water and the regime itself to these two riparians, it is noteworthy that the Indus Waters Treaty has survived two wars (one in 1965 and the other in 1971). In addition, the cooperation over water that has existed between the two riparians since the treaty's inception has been exceptionally stable and productive (Zawahri 2009).

Several empirical studies are likewise relevant to the discussion between scarcity and cooperation. Shira Yoffe, Aaron Wolf, and Mark Giordano (2003), for example, find no significant relationship between water scarcity and conflict events. Interestingly, they find only marginal positive effects between scarcity and cooperative events—although they based this particular claim on only eleven river basins. Other scholars focusing particularly on cooperation include Marit Brochmann and Nils Petter Gleditsch (2006), Brochmann and Paul Hensel (2009), and Jaroslav Tir and John Ackerman (2009). The first work operationalizes cooperation as membership in an international governmental organization (IGO) or dyadic trade. Scarcity is operationalized as the instance of drought in a given country in the past five years for the period 1975–2000. The authors find that when drought is interacted with a component used to depict whether the two countries actually share the river basin, cooperation in the sense of "additional dyadic trade" among the riparians is supported. Drought alone contributes to more IGO participation among dyads but not to more trade. Brochmann and Hensel (2009) also show that while water stress may lead to the onset of river claims among the protagonists, instances of negotiation and cooperation to settle these claims likewise increases. Tir and Ackerman (2009) uncover similar results between scarcity and water treaty formation, although they include other control variables to explain cooperation such as the type of political regime, the history of conflict between the parties, and the geographic configuration of the river.

Another recent study on the issue of flood control and institutional capacity argues that one of the reasons that the issue of floods is underrepresented in treaties may be due to the infrequency of them (Bakker 2009). In particular, the great majority of the forty-three river basins examined by the author experienced only one transboundary flood in the past twenty-one years. The relative absence of "scarcity in flood control," in other words, clarifies the lack of institutional capacity related to flood control.

The above analysis and accompanying studies imply that the relationship between scarcity, conflict, and cooperation is linear (or logarithmic). To put it another way, as water scarcity increases it is argued that the likelihood of cooperation increases as well. While Mark Giordano, Meredith Giordano, and Aaron Wolf (2005) have considered a nonlinear relationship between resource abundance in general and conflict, a similar relationship can be hypothesized for water scarcity in particular and cooperation. In other words, the relationship between water scarcity and formal cooperation, codified in an international water agreement, follows an inverted U-shaped curve (Dinar 2009). In essence, when natural and other resources are abundant, schemes of cooperation become superfluous. Conversely, when conditions are so harsh, fruitful ventures break down. A situation of moderate (or relative) scarcity therefore provides a suitable impetus for action between parties (Rawls 1971, 127–128; Ostrom et al. 1999, 281). By extension, if water was abundant, a treaty dividing the waters would be somewhat unnecessary. Instances of high scarcity would, conversely, also discourage cooperation. If water was extremely scant, the parties would have little to divide among themselves or would not be able to share any of the benefits that could be thereby derived.

Hydropower, flood control, and pollution issues should follow a similar pattern. For example, if little or no pollution exists in the river, there is little incentive to negotiate a pollution abatement agreement. At the same time, high pollution requires the riparians to exert sizable costs and efforts, which may be a deterrent to pollution abatement.[11] That being said, and in contrast to water quantity issues, cooperation over hydropower, flood control, and pollution should not be bound by the same fixed characteristics. Therefore, relative to water allocation issues, it is expected that coordination in the context of these other issues is more likely to take place where scarcity is high.

In an effort to empirically test this conjecture, another study (Dinar, Dinar, and Kurukulasuriya 2011) considers the corpus of rivers shared by only two states, evaluating dyads with and without formal treaties. While taking other control variables into account, this study finds a significant positive relationship between moderate scarcity levels and cooperation, and a negative relationship between high and low levels of water scarcity and cooperation. In other words, dyads with moderate scarcity levels exhibited a higher frequency of treaties, while dyads with low or high water scarcity exhibited little or no tendency toward formal cooperation. While this study includes the aforementioned issue areas (hydropower, pollution, and flood control) in its corpus of treaties, it employs a scarcity measure associated with water quantity issues (water availability per capita). The water quantity component of this analytic framework is thus of particular interest. While the study's findings are clearly in line with the emerging literature that highlights the relationship between scarcity and cooperation, it is the right side of its inverted U-shaped curve that calls for additional attention. That is, given high levels of scarcity, nations are less likely to conclude international water agreements. To that extent, external involvement (e.g., third parties) may be particularly salient in encouraging cooperation in these contexts. This may be especially relevant in Africa, which boasts the largest number of country dyads falling within these high-scarcity ranges yet lacking treaties (Dinar 2009).

Beyond Scarcity: The Importance of Other Factors for Explaining Conflict and Cooperation

While scarcity is important for understanding both conflict and cooperation, other factors are relevant as well. Scarcity may be a likely condition, in other words, but not sufficient to explain either conflict or cooperation over water.

Empirical studies such as those cited above have considered the extent of trade between the states, their overall historical relations, the nature of the riparians' political regimes, the viability of domestic institutions, the distribution of power, and the level of economic development in order to examine how these control variables help to explain conflict and cooperation over water.

More generally, this section considers several overarching factors that are important for explaining both conflict and cooperation over transboundary water in more detail. These include, the geographic characteristics of a river basin and the location of the riparians along the river, the relative distribution of power among the basin's riparians, and the domestic political arenas in the respective riparian countries.

Geography

The starting point for contemplating both conflict and cooperation over international water reposes in the river itself. The physical geography of the river defines the possibilities for *where*, *how*, and *when* the multiple uses of its water can be developed as well as used by riparian states (Elhance 1999, 15). In addition, the imposition of political boundaries on rivers creates different geographic relationships between basin countries, which often provide different incentives for cooperation.

According to David LeMarquand (1977, 8), for example, "successive" rivers (i.e., rivers where there is a clear upstream and downstream riparian) and "contiguous" ones (i.e., rivers where the river forms some part of the border between the two states) produce different incentives or disincentives for cooperation. When the river is contiguous there is significant incentive for cooperation. The incentive to attain such cooperation is to avoid the "tragedy of the commons" (LeMarquand 1977, 9). Alternatively, cooperation finds little incentive when the upstream country uses the river's water to the detriment of the downstream country, and that country has no reciprocal power over the upstream country (LeMarquand 1977, 10). As Peter Rogers (1993, 118) attests, the unidirectional feature of some rivers means that the resolution of river conflicts through the mutual control of reciprocally operating external effects (as in a contiguous geography) is generally ruled out.

Scholars have maintained that rivers, which are increasingly contiguous, are the hallmark of common property resources (Dasgupta, Mäler, and Vercelli 1997, 2). It is because all parties do not necessarily have to bear the full economic consequences of their actions that the successive configuration confers certain powers on the upstream country (Durth 1996, 62). Cooperation is therefore seldom attained due to the disincentives entrenched in this geographic configuration (Dolšak and Ostrom 2003, 347). While this scenario is applicable to several issues, water pollution concerns provide a most instructive case (Giordano 2003, 114).

A downstream nation will likely ask for strict controls of water pollution caused by its upstream nation. In turn, upstream states may be far less inclined to take the problem seriously, let alone bear the responsibility for devising an appropriate solution, as compared to the downstream state (Faure and Rubin 1993, 22–23; Matthew 1999, 171). Similarly, while water quality is of interest to downstream countries, the upstream countries have little incentive to invest in pollution prevention. In this context, the upstream state likes things the way they are and the downstream state wants to change them; effectively the latter party has less bargaining power than the former (Rangarajan 1985, 188). Hence, regime formation is unlikely to occur because the externalizer has no incentive to shoulder the burden of abating pollution (List and Rittberger 1992, 98, 101).

The situation regarding a contiguous river may be different, at least when the parties share similar shadows of the future. Pollution from wastewater effluents, for example, also affects the banks and territory of the country responsible for the pollution, just as it affects the neighboring state—and to that degree the externality is reciprocal. The incentive to abate pollution or prevent it before it is discharged into the water is thus intrinsic to the geography of the river. For this reason, pollution may be less of a problem for this kind of geography and may provide additional incentives for cooperation. As Scott Barrett (1994, 28) notes, reciprocal externalities differ from unidirectional ones in that a direct means exists by which one party may punish or reward the other's behavior. In fact, Hillary Sigman (2002a, 17), measuring pollution quantities along and across state borders in the United States, has shown that "water quality is significantly lower at stations upstream and downstream of state borders than at other stations," such as those along contiguous rivers. If pollution occurs, "the polluting country continues to experience damages for border rivers, so perhaps the incentives for control are sufficiently great to offset the lack of natural attenuation" (Sigman 2002a, 13). Sigman (2002b, 1157–1158) finds somewhat similar results for international rivers, at least in the case of upstream stations relative to other ones.

Relative Power

While LeMarquand's geographic analysis (1997) is generally deterministic, in considering cooperation less likely in upstream/downstream situations he maintains that cooperation in such scenarios is still possible when the downstream riparian has some reciprocal power over the upstream one. As Helga Haftendorn (2000, 63) notes, "Symmetry between the actors will only exist as far as the hydrological asymmetry between the actors is balanced out by other factors."

This scenario has been largely supported in the literature. Miriam Lowi (1993, 10), for example, has argued that in river basins where the upstream state is also the economic and military hegemon, cooperation is least likely to materialize since that state will have no obvious incentive to cooperate. According to Lowi, the case of the Euphrates-Tigris River best fits this criterion. Upstream Turkey is also the hegemon in comparison to midstream Syria and downstream Iraq. A comprehensive water agreement has not been forthcoming in the basin largely due to Turkey's intransigence. In short, Turkey's upstream and hegemonic position will continue to dissuade it from giving up the status quo that it benefits from to date (Lowi 1993, 10; Lowi 1995, 139).

Conversely, cooperation is most likely when the hegemon is downstream and if the relationship of the hegemon to the water source is that of critical need. The riparian, as a result, will "advocate or impose" a cooperative arrangement with its upstream neighbors (Lowi 1993, 10). Lowi (1993) claims that the Nile River Agreement (1959) between Egypt and Sudan is evidence of this latter hegemonic contention.[12]

Lowi's assertions afford two interesting and different implications for conflict and cooperation over water. First, upstream hegemons are less likely to cooperate since they are able to utilize the river's water unilaterally. Conflict of interest between the riparians over the river's utilization is thus likely to persist indefinitely. Second, cooperation is only likely to take place when the hegemon is downstream and perceives the water source to be that of a "critical need." In this instance the hegemon will assume the responsibility for initiating and enforcing the cooperative regime. Inferred is that the regime will be compelled and enforced by the hegemon (Lowi 1993, 11).[13]

While Lowi's theory (1993, 203) particularly concerns arid regions, her conclusion about upstream hegemons, and their adversity to cooperate, and downstream hegemons as malign actors in bringing about cooperation, is clearly suggested to be globally germane. Lowi (1993, 169) summarizes the predicament of the weaker state as follows:

> Since the asymmetry of power is not in its favor, it is not in a position to achieve its aims and satisfy its needs in an optimal fashion. Its capabilities are inferior to those of its adversary. In effect, it has little alternative but to accept a *modus vivendi* dictated by the stronger.

Domestic Politics

Domestic politics may also explain the inclination for conflict or cooperation. In essence, shifting and differing interests of domestic elements reflect how states formulate their initial position regarding a given issue (Iklé 1964, 122; Hopmann 1996, 155; Putnam 1998, 434). Applying this line of thinking to water issues, Fredrick Frey (1993, 63) has argued that "a nation's goals in transnational water relations are usually the result of internal power processes, which may produce a set of goals that does not display the coherence, transitivity or 'rationality' assumed in many analyses of transparent national interest." Particularly because water is considered a national security concern in some reaches of the world it may often be embroiled in nationalistic and identity issues (Elhance 1999). This overall claim has been generally supported by two main studies (Giordano, Giordano, and Wolf 2001; Pachova, Nakayama, and Jansky 2008). Both volumes find that international water conflict and cooperation are influenced by domestic water events, and vice versa.

In particular, hydropolitical relations between India and Bangladesh seem to be vulnerable to domestic political forces. Despite recent democratization, Bangladesh remains susceptible to Islamic fundamentalism and other domestic political factions that accuse their government of compromising the nation's sovereignty and national interest if they pursue negotiations with India. Such posturing often curtails cooperation and undercuts the would-be benefits (Elhance, 1999, 169–171).

Another example from Southern Asia is that of Nepal-India hydropolitics. In the future, the parties may seek to transcend their limited agreements for exploiting their shared waters.[14] A sector of Nepalese society nonetheless regards some of the water agreements already negotiated with India as providing that country with great benefits at the expense of Nepal. As such, desires for pursuing additional cooperation with India have been tainted with domestic skepticism, culminating in the incorporation of a clause in the country's constitution that requires any treaty pertaining to the exploitation of Nepal's natural resources to be ratified by the National Assembly by a two-thirds majority (Verghese 1996, 39–40; Shrestha and Singh 1996, 87; Gyawali 2000, 140). Future cooperation among the parties will therefore be subject to strong domestic scrutiny and final approval.

The above examples seem to support a general claim that domestic political support for hydropolitical cooperation is frequently hard to generate and sustain (Elhance 1999, 237). Yet examples can also be provided to demonstrate how domestic forces can play a positive role in facilitating cooperation.

One such instance is the dispute between the United States and Mexico over the Colorado River salinity problem. In the early 1960s, Mexico protested to the United States that the water it was entitled to under an earlier agreement—Treaty between the United States of America and Mexico (1944)—was highly saline. Despite the signing of interim solutions, the United States and Mexico agreed to a definitive solution by negotiating Minute 242 (1973). Essentially, the United States would construct a desalting plant that would deliver less saline water to Mexico. The desalting plant built in the United States was not one of the cheaper alternatives that could have been used to improve the water quality going into Mexico. Still, the desalting plant, funded by the federal government, not only satisfied the desires of all the Colorado River riparians within the United States, but also the concerned bureaucratic elements. As LeMarquand (1977, 44) has argued:

> The basin-wide salinity control program satisfied the Environment Protection Agency (EPA) by making its standards more politically acceptable; the Bureau of Reclamation . . . has found new opportunities to employ its talents in salinity control; the Department of the Interior now has the opportunity for the first time to implement the expertise and technological advances developed by its office of Saline Waters in a grand showpiece desalting plant; the lower basin states gain some assurances of a slower rate of increase in salinity concentration; the upper basin states will not have future water resource development curtailed.

Despite the negative effects of fundamental forces within Bangladesh on India-Bangladeshi hydropolitical relations, it is also instructive to note instances of cooperation in this particular arena as a function of domestic forces. The Ganges River Agreement (1977), for example, was largely a result of elections that took place in India (Elhance 1999, 177). Specifically, the Janata Party, which replaced the Congress Party in March 1977, not only wanted to usher in an era of improved relations with Bangladesh (which began to deteriorate in 1975), but was much more eager to reach an agreement over the Ganges, which was eluding the parties for quite some time (Salman and Uprety 2002, 151). The Ganges River Agreement (1996), a consequence of the various failures and limitations of the 1977 treaty as well as the subsequent memoranda of understanding, was likewise affected by domestic politics. In Bangladesh, the secular and center-left Awami League returned to power (Nishat and Faisal 2000, 300). In India, the United Front government took power in Delhi (Salman and Uprety 2002, 170). India's new foreign minister at the time, Inder Kumar Gujral, claimed that India needed to be more generous with its smaller neighbors. Known as the Gujral Doctrine, the principle was a major factor in shaping India's political and economic relations with Bangladesh (Salman and Uprety 2002, 170).

Bargaining Strategies and Strategic Interaction

History clearly demonstrates that the number of cooperative ventures over water surpasses the handful of military skirmishes that have taken place over water. Political disputes and conflicts of interest over shared water do indeed occur, yet the cooperative ventures and regimes that normally arise in response to these disagreements are most impressive.

Mutual interest and interdependence surely motivate river riparians to coordinate their actions, negotiate agreements, and cooperate. But the interdependence among the riparians may not always be equal (Keohane and Nye 1977, 9–11). In other words, scarcity may not always be mutual.

As suggested above, conflict may be exacerbated by the geographic asymmetries between states, but this asymmetry is not deterministic. Power asymmetries are not deterministic either. Contrary to the relative power argument, cooperation has also taken place in instances where the upstream states are the hegemons. Similarly, cooperative regimes that have taken shape in situations where the downstream state is the hegemon have not necessarily exhibited a coercive nature whereby the weaker state is at the mercy of the strong. In this context, countries with superior structural power find that they can achieve more by making promises and offering rewards, rather than by relying on threats and punishments (Young 1994, 135). These scenarios are depicted below through various cooperative examples pertaining to issues such as pollution, hydropower, flood control, and water allocation.

It is important to acknowledge that some scholars have considered the role of third parties and mediators in explaining hydropolitical cooperation (Browder 2000; Zawahri 2009). It is noteworthy, however, that many cases of interstate cooperation over water have culminated without the help of a mediator. Likewise, scholars have touted the role of international water law, especially as it pertains to the recent United Nations Convention on the Law of the Non-navigational Uses of International Watercourses (1997), in guiding states toward cooperation (Dellapenna 2001; McCaffrey 2001). For instance, scholars have noted how the convention's *equitable and reasonable utilization* principle has been applied in practice in the context of international water agreements in Africa (Lautze and Giordano 2006). In opposition, other analysts have argued that the convention has nonetheless stirred much controversy and dispute. The convention's ambiguity, for one, has been one of its main problems. Its perceived favoritism toward certain geographically situated states along a river as opposed to others has been another issue of concern to particular riparians (Schwabach 1998, 258).[15]

Despite the role of third parties or international legal principles, a review of international water treaties demonstrates that states have often been relatively successful in devising their own particular concomitants to property rights disputes. Asymmetries are usually offset via specific strategies such as side payments and issue linkage, which the protagonists frequently employ (either directly or indirectly) in their actual negotiations (LeMarquand 1977; Weinthal 2002; Dinar 2008).[16] These strategies are particularly effective as payoffs to cooperation are manipulated by the actors. Focusing on side payments, Barrett has claimed that treaties will be self-enforcing and stable if they are individually and collectively rational. In addition, such self-enforcing international treaties should have the effect of restructuring the incentives of the parties in order to conform their behavior to the treaties' tenets. Side payments may be one way of restructuring incentives and altering a state's behavior (Barrett 2003, 338–340, 351). Institutions and regimes are thus strengthened as compliance problems and cheating are reduced through these types of incentives and payoff-inducing strategies (Axelrod 1984, 13–31; Axelrod and Keohane 1985, 228, 231; Oye 1986, 15; Stein 1990, 184; Victor, Raustiala, and Skolnikoff 1998, 12).

Pollution

The utility of the above-mentioned bargaining strategies are most salient in the case of pollution control. While two states may share a river basin, a downstream state may be more dependent on the upstream state to abate pollution. Given the nature of this unidirectional externality, the downstream state will be more concerned, relative to the upstream state, about urgently dealing with the pollution (Matthew 1999, 171). The upstream state may therefore use its geographic upper hand to its strategic advantage, prolonging the dispute. That being said, both parties may still achieve a mutually satisfying outcome from cooperation. The downstream state, on the one hand, may guarantee abatement upstream, albeit at the cost of having to provide particular incentives to the upstream state. The upstream state, on the other hand, may gain both lucrative inducements and a form of recognition that it is not solely responsible for mitigating the harming activity since benefits follow pollution abatement.

These were the circumstances that set the context for the Rhine River Agreement (1976). Specifically, the Netherlands, located downstream, was the victim of chloride emissions (in other words, suffering from scarcity in pollution control or clean water) originating in upstream France and, to a lesser degree, Germany and Switzerland. In aggregate terms the French and Germans also constitute the more powerful riparians. Indeed, the Dutch endeavored to persuade the upstream states to abate their salt emissions, yet they in turn refused to underwrite all the abatement costs (LeMarquand 1977, 119; Barrett 2003, 128–132).

It was not until a compromise was reached that stipulated the Dutch contribution to the costs of abating pollution upstream that Germany and France were prompted to participate in a cost allocation regime and eventually cooperate. Interestingly, this solution did not include a bilateral exchange—that is, compensation from the Netherlands to France and Germany. Rather, a multilateral compensation regime was negotiated. As Thomas Bernauer (1996, 209–210) explains, by agreeing to contribute to the abatement costs in a multilateral setting (whereby other countries would pay the polluter as well), the Dutch played the other upstream countries against each other so that it did not have to pay all the abatement costs on its own.

Compensation also facilitated several agreements over pollution coming from Mexico into the United States, particularly over the Tijuana and New rivers.[17] In contrast to the Rhine River case, Mexico is located upstream on these two rivers. Additionally, not only is Mexico economically and militarily inferior to the United States but it also generally assumes a shorter shadow of the future with regard to the resource. Mexico's higher propensity to pollute relative to the ability of the United States to accept this pollution, in other words, is exacerbated by an upstream position. This makes the United States extremely dependent on Mexican actions vis-à-vis abatement on these two rivers.

Six agreements (or minutes) have been signed between the two riparians concerning specific pollution-related actions. Interestingly, the two initial agreements dealing with the Tijuana and New rivers called on Mexico to internalize the entire costs of abatement—Minute 270 (1985) and Minute 264 (1980), respectively. While Mexico took some actions as stipulated by these two agreements, it was quickly realized that continued pollution and sanitation problems coming from Mexico demanded a different strategy. In the case of both rivers, the United States had little choice but to contribute to the cost of abatement. Specific to the Tijuana, it was proposed that Mexico would not have to complete its undertakings from Minute 270. An international wastewater treatment plant would be built instead in the United States so as to treat sewage that would otherwise have continued to flow from Mexico into the United States—Minute 283 (1990) and Minute 296 (1997). Additional works were stipulated for the Mexican side, and the United States contributed to these costs as well—Minute 298 (1997). Specific to the New River, the two agreements that followed Minute 264 were in response to continued pollution from Mexico. Plants were to be built in both cases in Mexico in an effort to abate pollution—Minute 274 (1987) and Minute 294 (1995). The United States again contributed to the costs of these efforts.

The above cases demonstrate that downstream states often have to contribute to the costs of pollution control in geographically asymmetrical contexts. Thus, in contrast to the normatively accepted "polluter pays principle," states that are victims of pollution frequently have to contribute to the cleanup tab. Insisting that the polluting upstream state entirely pay for the pollution will normally not lead to immediate cooperation. Moreover, where the riparians share different shadows of the future with regard to the resource, pollution abatement will be increasingly urgent. Richer states generally may either have to provide some incentive to a poorer upstream state so as to abate pollution or absorb the financial burden for instituting increased pollution controls. The states' differing propensities to accept pollution and ability to pay for its abatement entail some form of a compensatory regime to solve the property rights dispute (Dinar 2006, 429).

Interestingly, there are few examples of specific agreements dealing with pollution control. Most agreements are general in that they codify either "indefinite commitments" or "defined activities" (Giordano 2003, 121–122). Nonetheless, the above cases may offer a precedent for other riparians suffering from specific pollution problems. An unresolved water quality dispute that comes to mind unfolds in the Kura-Araks Basin, primarily shared between Georgia, Armenia, and Azerbaijan. Azerbaijan, which is the farthest downstream, highly depends on the Kura-Araks water for its drinking supply. In turn, Kura-Araks water in Georgia and Armenia is mainly used for agriculture and industry (Vener and Campana 2010). Azerbaijan does not just suffer from water shortages, though; the water available for its use is highly polluted, in part due to upstream pollution in Georgia and Armenia. Various political issues (e.g., the Nagorno-Karabakh issue between Armenia and Azerbaijan) are currently retarding hydropolitical cooperation (Campana et al. 2008, 163). That being said, the use of bargaining strategies may prove useful when serious negotiations take place.

In particular, Azerbaijan may be in a position to use its significant oil and gas reserves to compensate Georgia as well as to some degree Armenia for pollution control, which is of less concern upstream. Issue linkage also may be employed with Azerbaijan to provide energy in exchange for guarantees of cleaner water. Both Georgia and Armenia require funding for domestic projects or lack energy-related natural resources, and hence could benefit from such a deal.

Hydropower and Flood Control

While utilizing a river for hydropower and flood-control benefits may provide more opportunities for mutual gain, as compared to pollution control, asymmetries may still have to be overcome to facilitate cooperation.

The Parana River Agreement (1973) between Brazil and Paraguay exemplifies a situation of a mightier upstream riparian in need of an accord with its weaker downstream riparian as it faced large energy deficits. In fact, since domestic energy production in Brazil was not sufficient, an international location—a stretch of the river shared with Paraguay—was deemed most efficient. A series of negotiations between the two countries led to the accord to jointly build the Guaira-Itaipu hydropower complex (Laborde 1999).

The agreement stipulated that the electricity output would be shared equally. Also agreed was that the country that did not require nor use a part of its allocation could sell that portion to the other party. While Brazil was critically dependent on its share, Paraguay did not have the domestic capacity to absorb such energy amounts. Indeed, the agreement offered Paraguay the opportunity to acquire a lucrative export commodity (Elhance 1999, 47). Because the agreement established Paraguay's equality with Brazil on water issues, it also mitigated some of the gross disparities between the larger riparian and its smaller neighbor in terms of "economic development, military capabilities and regional and international clout" that otherwise could have had a destabilizing effect on mutual relationships (Elhance 1999, 48–49).

It is important to note that despite the general benefits bestowed on Paraguay and Brazil, the major debate over the Guaira-Itaipu complex revolved around the relatively low price that Paraguay had negotiated in the original treaty. Consequently, in January 1985 the two parties signed several revisions to the treaty to cover matters of financial compensation. Paraguay's compensation thus increased (Seyler 1990). More recently, with the election of Fernando Lugo in Paraguay in April 2008, Paraguay has been demanding even higher compensation for power sales to Brazil. While Brazil has expressed reluctance to revise the treaty, its dependence on Paraguay's energy share is likely to lead to some changes.[18]

In all, Brazil's need for energy and lack of petrochemicals amplified the urgency to jointly develop the Guaira-Itaipu hydroelectric complex, which resulted in the agreement in 1973. In addition, Paraguay's bargaining power rested on the fact that it owned 50 percent of the waters needed to fuel the hydroelectric dams (Kempkey et al. 2008). Efficiently harnessing the river's hydropower potential could only be achieved by cooperating with its downstream neighbor, turning Brazil into a benevolent upstream hegemon.

In short, the benefits of joint action promoted cooperation. Paraguay also gained compensation (and royalties) from the energy sales. Brazil's overall dependence on hydroelectric energy may induce a future renegotiation of the agreement.

The Columbia River Agreement (1961) stipulated an accord on both hydropower and flood control. While both Canada (the upstream riparian) and the United States (the downstream riparian) saw the benefits in utilizing the river for hydropower and flood-control purposes, it was the latter riparian that was in greater need of energy and flood protection (Federal Columbia River System 2003, 3).[19] Consequently, the geographic asymmetry in Canada's favor combined with the greater need of the United States for the agreement had to be offset to encourage cooperation.

Canada, unsurprisingly, was reluctant in the first place to go ahead with any projects unless it was assured of receiving some compensation for the unrealized benefits it was to send downstream to the United States (Lepawsky 1963, 542; Krutilla 1967, 10). Barrett notes that the United States believed that Canada would want to develop the Columbia River on its side of the border anyway, and so felt that it did not need to compensate Canada for constructing the project. When Canada threatened to build an alternative project on a different river that would provide the United States with no benefits, the United States heeded the threat as a credible one, and Canada was able to secure a more attractive deal (Barrett 1994, 22).

The agreement stipulated that Canada construct the dams, which would provide improved stream flow and regulation, and thereby increase hydropower generation (and flood control) in the United States. The United States paid Canada for constructing and maintaining the reservoirs, particularly pertaining to the equivalent storage capacity required by the United States. Payments were also given to Canada for operating those storage facilities in flood periods. Half the downstream power benefits produced in the United States and created by the enhanced water flow were also to be supplied to Canada. The Canadian entitlement was considered surplus energy at the time and was sold to the United States.

John Krutilla observes that the treaty cannot be considered an isolated affair. In addition to the compensation provided to Canada, issue linkage was affected too. According to Krutilla (1966, 96), the Columbia River system was an arena in which the United States could make an attractive arrangement in exchange for concessions, perhaps involving North American continental defense or other areas in which the vital interests of the United States were at stake.

In all, Canada's bargaining position was strengthened given its upstream position, where the majority of infrastructure per the Columbia River Agreement was to be built. The United States thus compensated Canada in order to facilitate cooperation.

The two cases presented above demonstrate that even with regard to positive externalities (i.e., the benefits that may be created from hydropower and flood-control works), geographic asymmetries may still have to be overcome to foster cooperation. Mutual gains from cooperation may also be unequal, and this may offer additional reasons for employing incentives. In the case of the Guaira-Itaipu hydropower plant, the project was constructed on the part of the river shared between the two riparians. Hence, there was an element of geographic symmetry that facilitated cooperation. Yet Brazil's greater need for the energy and Paraguay's acquiescence translated into a treaty that wrought benefits on the weaker downstream state as well—mostly in the form of revenues from energy sales.

Water Allocation

Like pollution issues, water allocation issues may also be subject to various asymmetries and strategic behavior. In principle, the upstream state may impound the water or prevent it from flowing downstream. This does not mean, however, that downstream states are unable to use particular strategies to offset such asymmetry.

The Tigris-Euphrates, a shared basin between upstream Turkey, midstream Syria, and downstream Iraq, is useful in demonstrating how bargaining strategies can be used to foster cooperation. While a comprehensive agreement between the three riparians is indeed absent, largely due to Turkey's intransigence as the hegemonic riparian, bilateral agreements stipulated between these riparians are noteworthy. Of the bilateral agreements that have been negotiated in the basin, the Protocol on Matters Pertaining to Economic Cooperation between the Republic of Turkey and the Syrian Arab Republic (1987) over the Euphrates is most revealing for asymmetrical bargaining contexts, since issue linkage was the primary tactic used by the downstream and relatively weaker state.[20]

The protocol stipulated that Turkey guarantee Syria a minimum flow of 500 cubic meters per second (or about 16 billion cubic meters a year) in the Euphrates River. The agreement also declared that if the monthly flow should fall below that level, Turkey will make up the difference during the following month. This amount was well within the range demanded by Syria in earlier negotiations (Elhance 1999, 144). In return for this guaranteed flow, Syria made concessions on border issues that ranged from the smuggling of illegal arms and narcotics to infiltration into Turkey by separatist groups, primarily the Kurdish Workers' Party (Daoudy 2009). In 1989, as a result, Syria signed an accord with Iraq by which it would retain 42 percent and release 58 percent of the annual flow it receives from the Euphrates to Iraq.[21]

Since 1987, Syria (and Iraq to some degree) has continued to use linkage strategies to offset Turkey's powerful stance in the basin. Between 1993 and 2002, for example, Syria blocked international investment in Turkey's upstream projects (particularly its Southeastern Anatolia Project [or GAP, in Turkish]) by appealing to European export credit agencies and the World Bank. One of the noteworthy results has been the withdrawal of British and Swiss investors from GAP-related projects (Daoudy 2009). Interestingly, relations between Turkey and Syria have progressed since the second Gulf War. While an agreement on equitable water sharing is still absent, the two states are working closely on matters relating to benefit sharing and joint projects (Daoudy 2008, 99–100). Meetings have also been held among all three riparians in an effort to resume trilateral water talks (Daoudy 2008, 91).

The Helmand River Agreement of 1973 is another example pointing to the use of both side payments and issue linkage. In this particular instance, these strategies were used by the stronger downstream riparian, Iran, to guarantee additional allocations of water from upstream Afghanistan. A. H. Abidi (1977) provides a specific account of the negotiation process preceding the treaty. Accordingly, the agreement called on Afghanistan to release a set amount of water a year for Iran's use from the Helmand River. As Abidi (1977, 370) adds, "Iran offered financial payment and concessional transit rights for Afghan exports through Bandar Abbas in return for more water by Afghanistan."

As opposed to the benefits that may be created via pollution control, hydropower generation, and flood prevention, and the compensation regime that ensues, side payments are rarely used in the realm of solving water allocation disputes. This is largely a function of modern international principles such as equitable utilization, which undergirds the claim that states don't solely own the waters of an international river flowing in their territory. Upper riparians are under an obligation not to prevent such waters from flowing downstream (McCaffrey 2001, 264). Issue linkage, though, is a common strategy for facilitating water allocation treaties. Other examples include the Treaty between the United States of America and Mexico (1944) along with the various agreements over the Syr Darya and Amu Darya rivers in Central Asia.[22] The former regime witnessed Mexico linking the two rivers in a bid to use its upstream position on the Lower Rio Grande tributaries so as to gain additional water allocations on the Colorado, where it was completely downstream (Bennett, Ragland, and Yolles 1998, 67; Fischhendler and Feitelson 2003). The regimes pertaining to the Syr Darya and Amu Darya exhibit the exchange of fuel from downstream states such as Uzbekistan and Kazakhstan for seasonal water releases from upstream states such as Kyrgyzstan and Tajikistan.

Conclusion

In general, the history of hydropolitics is one of negotiation and cooperation rather than militarized conflict. This does not mean that political conflicts and disputes over water do not take place. Yet such conflicts are frequently resolved through international water agreements. While water scarcity may provide the impetus for conflicts between states, it oftentimes affords the motivation for cooperation. While some studies have demonstrated that water scarcity and cooperation generally exhibit a linear relationship, others hold that cooperation in the form of international water agreements is more likely when scarcity is neither too high nor too low among the parties. In other words, formal cooperation is anticipated in times of relative or moderate scarcity. These latter findings indicate that cooperation may be impeded when scarcity is too severe, implying that exogenous factors may well be necessary to facilitate interstate cooperation. Despite the nuances of the relationship between scarcity and cooperation, however, the overall findings offer a strong challenge to the popular prediction that "the wars of the next century will be about water."[23]

Surely cooperation can't solely be explained by scarcity and the desire to harness the river for mutual gain. Other overarching variables facilitate (or impede) negotiations between states. Considering the role of such variables as geography and relative power, this chapter argues that it is the asymmetries embedded in international negotiations over water that usually need to be offset to encourage cooperation.

Strategies such as issue linkage and side payments are of paramount importance in counterbalancing asymmetries as states attempt to manipulate the payoffs to cooperation of their neighboring riparians (those states perhaps less pressed for an agreement). Several examples, in the form of international water treaties, are provided across the various nonnavigational and nonfishery water scarcity issues discussed in the chapter, including water quality, hydropower and flood control, and water allocation. This type of analysis is not necessarily interested in the quality of these treaties or their particular assessment but rather the general mechanisms used to bring about (and sustain) such water regimes.

Consequently, in the geographic context alleged to be most prone to continued dispute—upstream/downstream configuration—the chapter shows that cooperation can nonetheless be sustained. This is the case even if the upstream riparian is the hegemon and in principle may have the least interest in a cooperative regime since the immediate economic incentives to cooperate are not clear.

Such an analysis also points to the fact that in the realm of hydropolitics, downstream hegemons often play a benign role, providing their upstream neighbors with incentives to cooperate as they are dependent on their acquiescence and require their strategic territory to harness the river.

Both aforementioned scenarios, depicting two different geographic and power-based settings, also demonstrate that scarcity does not have to be mutual for cooperation to take place. As evidenced in the pollution cases, one party (generally the downstream state) may be more urgently seeking an agreement to foster upstream pollution abatement. This inherent asymmetry, exacerbated by the geographic and economic discrepancies in the basin, can be offset in an effort to facilitate cooperation.

In all, if asymmetries in hydropolitics present some of the major obstacles to cooperation, the examples here illustrate that through a set of incentives, payoff structures may be manipulated so as to encourage regime creation.

Notes

1. Leo Lewis, "Water shortages are likely to be trigger for wars, says UN chief Ban Ki Moon," *The TimesOnline*, December 4, 2007. Available at <http://www.timesonline.co.uk/tol/news/world/asia/article2994650.ece> (accessed February 5, 2008).

2. Ben Russell and Nigel Morris, "Armed Forces Are Put on Standby to Tackle Threat of Wars over Water," *Independent*, February 28, 2006, available at <http://www.independent.co.uk/environment/armed-forces-are-put-on-standby-to-tackle-threat-of-wars-over-water-467974.html> (accessed February 5, 2008).

3. Thomas Homer-Dixon (1999, 139) claims that the incentives to launch wars over renewable resources are low. The exception may be water, however. Specifically, such resource wars are likely in upstream and downstream situations when the downstream state fears that the upstream one will use water as a form of leverage, and believes it has the military power to rectify the situation. Michael Klare (2001, 60) argues that while the actual use of force in resolving water disputes has been relatively rare, the growing pressure on vital supplies will create more frequent clashes. For other policy-oriented writings often associated with the water wars thesis, see Cooley 1984; Starr 1991; Bullock and Darwish 1993.

4. See also Toset, Gleditsch, and Hegre 2000; Pacific Institute n.d.. Some analysts have also regarded the dispute over water installations between Syria and Israel, given the former's efforts to divert the Jordan River's headwaters, as a catalyst for the 1967 War (Cooley 1984; Westing 1986; Falkenmark 1986; Myers 1993, 38). Rejecting this "hydraulic imperative" theory, Wolf (1998, 254) claims that while "shots were fired over water between Israel and Syria from 1951 to 1953 and from 1964 to 1966, the final exchange, including both tanks and aircraft on July 14, 1966, stopped Syrian construction of the diversion project in dispute, effectively ending water-related tensions between the two states. The 1967 war broke out almost a year later."

5. FAO 1978/1984; Teclaff 1967; Oregon State University n.d.; Dinar 2008.

6. For other academic writings that cast doubt on the water wars thesis, see Allan 2001; Deudney 1999; Ohlsson and Turton 2000; Ohlsson 1999. Acknowledging that water is a source of cooperation, an article posted on the prominent Environmental Change and Security Program Web site and associated with the program's water initiative interestingly notes that "this [cooperative] history does not, however, close the door on the international 'water war' debate for the present or in the future. Because as many as 7 billion people, more than currently alive in the world today, may live under conditions of water scarcity and stress in 2050, the future may not resemble the past when it comes to violent conflict between states" (Working Group II of the Navigating Peace Initiative n.d.). Given

the effects of climate change on water resources, a recent U.S. intelligence report observes, "Climate change could threaten domestic stability in some states, potentially contributing to intra- or, less likely, interstate conflict, particularly over access to increasingly scarce water resources" (House Permanent Committee on Intelligence 2008, 5).

7. The literature on conflict and cooperation over water is too vast to cite here, but some examples include Naff and Matson 1984; Gleick 1993; Rogers and Lydon 1994; Scheumann and Schiffler 1998; Beach et al. 2000; Lonergan 2001; Blatter and Ingram 2001; Kibaroğlu 2002; Turton and Henwood 2002; Dombrowski 2007; UNESCO n.d.. Additional examples are discussed throughout this chapter.

8. For a critique of the general utility of interstate regimes in governing water, see Conca 2006.

9. Malin Falkenmark (1986) contends that a thousand cubic meters of water per capita per year constitutes the minimum necessary for an adequate quality of life in a moderately developed country. When water availability drops below this figure, scarcity problems grow intense.

10. An important caveat to this example is the role of the World Bank in facilitating the agreement by providing the needed monies to implement the agreement. Also, the agreement stipulated that the waters of the Indus would be physically divided (in essence, particular tributaries would go to one riparian and the remaining tributaries would go to the other riparian) rather than be comanaged.

11. This might be especially salient among developing countries (Matthew 1999, 172).

12. See also Naff and Frey 1985; Naff 1994; Homer-Dixon 1999, 139–141. All three readings argue that when the basin's hegemon is upstream, conflict is least likely. This does not necessarily contradict Lowi's assessment regarding upstream hegemons, for these studies are only asserting that violent conflict (such as war), rather than a conflict of interest or political dispute, is not likely because the hegemon holds a large degree of power, which other states are too weak to challenge. On the other hand, the three works contend that violent conflict is most likely when the downstream state is also the basin's hegemon. While this claim seems to directly oppose Lowi's that cooperation is most likely when the hegemon is downstream, her logic for explaining the emergence of a cooperative regime stems from the potential for conflict in this type of basin power configuration and the need to confine any upstream riparian to the hegemon's interests.

13. For an assessment of the realist conception of cooperation, see Haas 1990.

14. See, for example, the Kosi River Agreement (1954) and the Gandak River Agreement (1959).

15. As of May 2010, only eighteen of the thirty-five countries needed for the convention to enter into force have signed, ratified, accepted, approved, or acceded to it. See <http://treaties.un.org/Pages/ViewDetails.aspx?src=UNTSONLINE&tabid=2&mtdsg_no=XXVII-12&chapter=27&lang=en#Participants> (accessed May 16, 2010). The deadline for ratification has long passed.

16. Other studies include Bennett, Ragland, and Yolles 1998; Fischhendler and Feitelson 2003.

17. For a detailed discussion of the Tijuana case, see Fischhendler 2007.

18. Alexei Barrionuevo, "For Former Bishop Turned President of Paraguay, a Difficult Road Ahead," *New York Times*, August 16, 2008, A6. Available at <http://www.nytimes.com/2008/08/16/world/americas/16paraguay.html?_r=1> (accessed February 5, 2008).

19. The storage facilities stipulated in the treaty have made substantial improvements in meeting the flood management objectives in the United States. In 1972 and 1974, which saw particularly high water discharges, the storage projects were credited with preventing losses of US$474 and US$538 million, respectively (Yu 2008, 42).

20. The Treaty of Friendship and Good Neighborly Relations between Turkey and Iraq, signed in 1946, is another example of a bilateral agreement in the region. The agreement underlined the positive impact of storage facilities (on Iraq) to be situated in Turkey in return for compensation from Iraq.

21. Law no. 14, ratifying the Joint Minutes concerning the provisional division of the waters of the Euphrates River, April 17, 1989.

22. In the case of Central Asia, see Agreement between the Government of the Republic of Uzbekistan and the Government of the Republic of Tajikistan (2000); Agreement between the Government of the Republic of Kazakhstan and the Government of the Kyrgyz Republic (1999); Agreement between the Governments of the Republic of Kazakhstan, the Kyrgyz Republic, and the Republic of Uzbekistan (1998).

23. This prediction, made by Ismail Serageldin, the Vice President of the World Bank at the time, is quoted in Barbara Crosette, "Severe Water Crisis Ahead for Poorest Nations in the Next Two Decades," *New York Times*, August 10, 1995, sec. 1, 13; Kofi Annan, the UN secretary general at the time, made a similar prediction in remarks addressed to the Association of American Geographers (2001) at its ninety-seventh annual meeting.

References

Abidi, A. H. 1977. "Irano-Afghan dispute over the Helmand waters." *International Studies* 16 (3): 357–378.

Agreement between the Government of the Republic of Kazakhstan and the Government of the Kyrgyz Republic on Comprehensive Use of Water and Energy Resources of the Naryn Syr Darya Cascade Reservoirs. 1999. May 22. Available at <http://www.ce.utexas.edu/prof/mckinney/papers/aral/agreements/Annual-KzKg-99.pdf> (accessed March 4, 2008).

Agreement between the Governments of the Republic of Kazakhstan, the Kyrgyz Republic, and the Republic of Uzbekistan on the Use of Water and Energy Resources of the Syr Darya Basin. 1998. March 17. Available at <http://ocid.nacse.org/qml/research/tfdd/toTFDDdocs/194ENG.htm> (accessed March 4, 2008).

Agreement between the Government of the Republic of Uzbekistan and the Government of the Republic of Tajikistan on Cooperation in the Area of Rational Water and Energy Uses. 2000. January 14. Available at <http://www.ce.utexas.edu/prof/mckinney/papers/aral/agreements/Kayrakum-00.pdf> (accessed March 4, 2008).

Alam, Undala. 2002. "Questioning the water wars rationale: A case study of the Indus waters." *Geographical Journal* 168 (4): 341–353.

Allan, Anthony. 2001. *The Middle East Water Question*. London: I. B. Tauris Publishers.

Association of American Geographers. 2001. "United Nations secretary general Kofi Annan addresses the 97th annual meeting of the Association of American Geographers." News archives. Available at <http://www.aag.org/news/kofi.html> (accessed October 5, 2006).

Axelrod, Robert. 1984. *The Evolution of Cooperation*. New York: Basic Books.

Axelrod, Robert, and Robert Keohane. 1985. "Achieving cooperation under anarchy: Strategies and institutions." *World Politics* 38 (1): 226–254.

Bakker, Marloes. 2009. "Transboundary river floods and institutional capacity." *Journal of the American Water Resources Association* 45 (3): 553–566.

Barrett, Scott. 1994. "Conflict and cooperation in managing international water resources." World Bank Policy Research Working Paper 1303. Washington, DC: World Bank.

Barrett, Scott. 2003. *Environment and Statecraft: The Strategy of Environmental Treaty-Making*. Oxford: Oxford University Press.

Beach, Heather, Hamner Jesse, Joseph Hewitt, Edy Kaufman, Anja Kurki, Joe Oppenheimer, and Aaron Wolf. 2000. *Transboundary Freshwater Dispute Resolution: Theory, Practice, and Annotated References*. Tokyo: United Nations University Press.

Bennett, Lynne, Shannon Ragland, and Peter Yolles. 1998. "Facilitating international agreements through an interconnected game approach: The case of river basins." In *Conflict and Cooperation on Trans-boundary Water Resources*, ed. Richard Just and Sinaia Netanyahu, 61–85. Boston: Kluwer Academic.

Bernauer, Thomas. 1996. "Protecting the river Rhine against chloride pollution." In *Institutions for the Earth: Pitfalls and Promise*, ed. Robert Keohane and Marc Levy, 201–232. Cambridge, MA: MIT Press.

Blatter, Joachim, and Helen Ingram, eds. 2001. *Reflections on Water: New Approaches to Transboundary Conflicts and Cooperation*. Cambridge, MA: MIT Press.

Brochmann, Marit, and Nils Petter Gleditsch. 2006. "Shared rivers and international cooperation." Paper presented at the International Studies Association Annual Meeting, San Diego, March 22–25.

Brochmann, Marit, and Paul Hensel. 2009. "Peaceful management of international river claims." *International Negotiation* 14 (2): 393–418.

Browder, Greg. 2000. "An analysis of the negotiations for the 1995 Mekong Agreement." *International Negotiation* 5 (2): 237–261.

Bullock, John, and Adel Darwish. 1993. *Water Wars: Coming Conflicts in the Middle East*. London: St. Dedmundsbury Press.

Burton, John. 1972. *World Society*. Cambridge: Cambridge University Press.

Campana, Michael, Berrin Vener, Nodar Kekelidze, Bahruz Suleymanov, and Armen Saghatelyan. 2008. "Science for peace: Monitoring water quality and quantity in the Kura-Araks Basin of the South Caucasus." In *Transboundary Water Resources: A Foundation for Regional Stability in Central Asia*, ed. John Moerlins, Mikhail Khankhasayev, Steven Leitman, and Ernazar Mkhmudov, 153–170. Dordrecht: Springer.

Columbia River Agreement (Treaty between the United States of America and Canada Relating to Cooperative Development of the Water Resources of the Columbia River Basin). 1961. January 17. Available at <http://ocid.nacse.org/cgi-bin/qml/tfdd/treaties.qml?qml_screen=full&TN=116> (accessed February 14, 2007).

Conca, Ken. 2006. *Governing Water: Contentious Transnational Politics and Global Institution Building*. Cambridge, MA: MIT Press.

Cooley, John. 1984. "The war over water." *Foreign Policy* 54: 3–26.

Daoudy, Marwa. 2008. "Hydro-hegemony and international water law: Laying claims to water rights." *Water Policy* 10 (S2): 89–102.

Daoudy, Marwa. 2009. "Asymmetric power: Negotiating water in the Euphrates and Tigris." *International Negotiation* 14 (2): 361–391.

Dasgupta, Partha, Karl-Göran Mäler, and Alessandro Vercelli. 1997. "Introduction." In *The Economics of Transnational Commons*, ed. Partha Dasgupta, Karl-Göran Mäler, and Alessandro Vercelli, 1–16. Oxford: Clarendon Press.

Dellapenna, Joseph. 2001. "The customary international law of transboundary fresh waters." *International Journal of Global Environmental Issues* 1 (3–4): 264–305.

Deudney, Daniel. 1999. "Environmental security: A critique." In *Contested Grounds: Security and Conflict in the New Environmental Politics*, ed. Daniel Deudney and Richard Matthew, 187–223. Albany: State University of New York Press.

Dinar, Shlomi. 2006. "Assessing side-payment and cost-sharing patterns: The geographic and economic connection." *Political Geography* 25 (4): 412–437.

Dinar, Shlomi. 2008. *International Water Treaties: Negotiation and Cooperation along Transboundary Rivers*. London: Routledge.

Dinar, Shlomi. 2009. "Scarcity and cooperation along international rivers." *Global Environmental Politics* 9 (1): 107–133.

Dinar, Shlomi, Ariel Dinar, and Pradeep Kurukulasuriya. 2011. "Scarcity and cooperation along international rivers: An empirical assessment of bilateral treaties." *International Studies Quarterly* (forthcoming in September).

Dokken, Karen. 1997. "Environmental conflict and international integration." In *Conflict and the Environment*, ed. Nils Petter Gleditsch, 519–534. Dordrecht: Kluwer Academic Publishers.

Dolšak, Nives, and Elinor Ostrom. 2003. "The challenges of the commons." In *The Commons in the New Millennium: Challenges and Adaptation*, ed. Nives Dolšak and Elinor Ostrom, 3–34. Cambridge, MA: MIT Press.

Dombrowski, Ines. 2007. *Conflict, Cooperation, and Institutions in International Water Management*. Cheltenham, UK: Edward Elgar.

Durth, Rainer. 1996. "European experience in the solution of cross-border environmental problems." *Inter Economics* 32: 62–67.

Elhance, Arun. 1999. *Hydropolitics in the 3rd World: Conflict and Cooperation in International River Basins*. Washington, DC: United States Institute of Peace Press.

Falkenmark, Malin. 1986. "Fresh waters as a factor in strategic policy and action." In *Global Resources and International Conflict: Environmental Action in Strategic Policy and Option*, ed. Arthur Westing, 85–113. Oxford: Oxford University Press.

Falkenmark, Malin. 1992. "Water scarcity generates environmental stress and potential conflicts." In *Water, Development, and the Environment*, ed. William James and Janusz Niemczynowicz, 279–294. Boca Raton, FL: Lewis Publishers.

FAO (Food and Agriculture Organization). 1978/1984. *Systematic Index of International Water Resources, Treaties, Declarations, and Cases by Basin*. Rome: Food and Agriculture Organization.

Faure, Guy Olivier, and Jeffery Rubin. 1993. "Organizing concepts and questions." In *International Environmental Negotiation*, ed. Gunnar Sjöstedt, 17–26. Newbury Park, CA: Sage Publications.

Federal Columbia River System. 2003 Available at <http://www.bpa.gov/power/pg/fcrps_brochure_17x11.pdf> (accessed February 4, 2007).

Fischhendler, Itay. 2007. "Escaping the 'polluter pays' trap: Financing wastewater treatment on the Tijuana–San Diego border." *Ecological Economics* 63 (2–3): 485–498.

Fischhendler, Itay, and Eran Feitelson. 2003. "Spatial adjustment as a mechanism for resolving river basin conflicts: The US-Mexico case." *Political Geography* 22 (5): 557–583.

Frey, Fredrick. 1993. "The political context of conflict and cooperation over international river basins." *Water International* 18 (1):54–68.

Giordano, Mark, Meredith Giordano, and Aaron Wolf. 2005. "International resource conflict and mitigation." *Journal of Peace Research* 24 (1): 47–65.

Gandak River Agreement (Agreement between His Majesty's Government of Nepal and the Government of India on the Gandak Irrigation and Power Project). 1959. December 4. Available at <http://ocid.nacse.org/tfdd/tfdddocs/111ENG.htm> (accessed April 15, 2008).

Ganges River Agreement (Agreement between the Government of the People's Republic of Bangladesh and the Government of the Republic of India on Sharing of the Ganges Waters at Farakka and on Augmenting Its Flows). 1977. November 5. Available at <http://ocid.nacse.org/tfdd/tfdddocs/158ENG.htm> (accessed August 22, 2008).

Ganges River Agreement (Treaty between the Government of the Republic of India and the Government of the People's Republic of Bangladesh on Sharing the Ganges Waters at Farakka). 1996. December 12. Available at <http://ocid.nacse.org/tfdd/tfdddocs/172ENG.htm> (accessed August 22, 2008).

Giordano, Mark, Meredith Giordano, and Aaron Wolf. 2005. "International resource conflict and mitigation." *Journal of Peace Research* 24 (1): 47–65.

Giordano, Meredith. 2003. "Managing the quality of international rivers: Global principles and basin practice." *Natural Resources Journal* 43 (1): 111–136.

Giordano, Meredith, Mark Giordano, and Aaron Wolf. 2001. "The geography of water conflict and cooperation: Internal pressures and international manifestations." *Geographical Journal* 168 (4): 293–312.

Gleditsch, Nils Petter. 1998. "Armed conflict and the environment: A critique of the literature." *Journal of Peace Research* 35 (3): 381–400.

Gleditsch, Nils Petter, Kathryn Furlong, Håvard Hegre, Bethany Lacina, and Taylor Owen. 2006. "Conflict over shared rivers: Resource scarcity or fuzzy boundaries." *Political Geography* 25 (4): 361–382.

Gleick, Peter, ed. 1993. *Water in Crisis: A Guide to the World's Freshwater Resources*. New York: Oxford University Press.

Gyawali, Dipak. 2000. "Nepali-India water resource relations." In *Power and Negotiation*, ed. William Zartman and Jeffrey Rubin, 129–154. Ann Arbor: University of Michigan Press.

Haas, Peter. 1990. *Saving the Mediterranean: The Politics of International Environmental Cooperation*. New York: Columbia University Press.

Haftendorn, Helga. 2000. "Water and international conflict." *Third World Quarterly* 21 (1): 51–68.

Hensel, Paul, Sara McLaughlin Mitchell, and Thomas Sowers. 2006. "Conflict management of riparian disputes." *Political Geography* 25 (4): 383–411.

Homer-Dixon, Thomas. 1999. *Environment, Scarcity, and Violence*. Princeton, NJ: Princeton University Press.

Hopmann, Terrence. 1996. *The Negotiation Process and the Resolution of International Conflicts*. Columbia: University of South Carolina Press.

House Permanent Committee on Intelligence. 2008. "National intelligence assessment on the national security implications of global climate change to 2030." June 25. Available at <http://www.dni.gov/testimonies/20080625_testimony.pdf> (accessed June 27, 2009).

Iklé, Fred. 1964. *How Nations Negotiate*. New York: Harper and Row.

Indus River Agreement (Indus Water Treaty between the Government of India, the Government of Pakistan, and the International Bank for Reconstruction and

Development). 1960. September 19. Available at <http://ocid.nacse.org/qml/research/tfdd/toTFDDdocs/114ENG.htm> (accessed October 22, 2007).

Johnston Negotiations. 1955. Available at <http://ocid.nacse.org/tfdd/tfdddocs/92ENG.htm> (accessed February 5, 2008).

Kempkey, Natalie, Margaret Pinard, Víctor Pochat, and Ariel Dinar. 2009. "Negotiations over water and other natural resources in the La Plata River Basin: A model for other transboundary basins." *International Negotiation* 14 (2): 252–279.

Keohane, Robert. 1982. "The demand for international regimes." *International Organization* 36 (2): 325–355.

Keohane, Robert, and Joseph Nye. 1977. *Power and Interdependence: World Politics in Transition.* Boston: Little, Brown.

Kibaroğlu, Ayşegül. 2002. *Building a Regime for the Waters of the Euphrates-Tigris River Basin.* London: Kluwer Law International.

Klare, Michael. 2001. "The new geography of conflict." *Foreign Affairs* 80 (3): 49–61.

Kosi River Agreement (Agreement between the Government of India and the Government of Nepal on the Kosi Project). 1954. April 25. Available at <http://ocid.nacse.org/tfdd/tfdddocs/85ENG.htm> (accessed April 15, 2008).

Kratz, Stefan. 1996. "International conflict over water resources: A syndrome approach." *Wissenschaftszentrum Berlin Fur Sozialforschung*, FS II, 96–401. Berlin: Social Science Research Center.

Krutilla, John. 1966. "The international Columbia River Treaty: An economic evaluation." In *Water Research*, ed. Allen Kneese and Stephen Smith, 69–97. Baltimore: Johns Hopkins University Press.

Krutilla, John. 1967. *The Columbia River Treaty: The Economics of an International River Basin Development.* Baltimore: Johns Hopkins University Press.

Laborde, Lilian del Castillo. 1999. "The Plata Basin institutional framework." In *Management of Latin American River Basins: Amazon, Plata, and São Francisco*, ed. Asit Biswas, Newton Cordeiro, Benedito Braga, and Cecilia Tortajada, 175–204. Tokyo: United Nations University Press.

Lautze, Jonathan, and Mark Giordano. 2006. "Equity in transboundary water law: Valuable paradigm or merely semantics." *Colorado Journal of International Environmental Law and Policy* 17 (1): 89–122.

LeMarquand, David. 1977. *International Rivers: The Politics of Cooperation.* Vancouver: Westwater Research Center.

Lepawsky, Albert. 1963. "International development of river resources." *International Affairs* 39 (4): 533–550.

List, Martin, and Volker Rittberger. 1992. "Regime theory and international environmental management." In *The International Politics of the Environment: Actors, Interests, and Institutions*, ed. Andrew Hurrell and Benedict Kingsbury, 85–109. Oxford: Clarendon Press.

Lonergan, Steven. 2001. "Water and conflict: Rhetoric and reality." In *Environmental Conflict*, ed. Paul Diehl and Nils Petter Gleditsch, 109–124. Boulder, CO: Westview Press.

Lowi, Miriam. 1993. *Water and Power: The Politics of a Scarce Resource in the Jordan River Basin*. Cambridge: Cambridge University Press.

Lowi, Miriam. 1995. "Rivers of conflict, rivers of peace." *Journal of International Affairs* 49 (1): 123–145.

Matthew, Richard. 1999. "Scarcity and security: A common-pool resource perspective." In *Anarchy and the Environment: The International Relations of Common Pool Resources*, ed. Samuel Barkin and George Shmabaugh, 155–175. Albany: State University of New York Press.

McCaffrey, Stephen. 2001. *The Law of International Watercourses: Non-Navigational Uses*. Oxford: Oxford University Press.

Minute 242, Agreement Setting Forth a Permanent and Definitive Solution to the International Problem of the Salinity of the Colorado River. 1973. August 30. Available at <http://ocid.nacse.org/qml/research/tfdd/toTFDDdocs/315ENG.htm> (accessed October 18, 2007).

Minute 264, Recommendation for Solution of the New River Border Sanitation Problem at Calexico, California-Mexicali, Baja California Norte. 1980. August 26. Available at <http://www.ibwc.state.gov/Files/Minutes/Min264.pdf> (accessed June 25, 2008).

Minute 270, Recommendations for the First Stage Treatment and Disposal Facilities for the Solution of the Border Sanitation Problem at San Diego, California-Tijuana, Baja California. 1985. April 30. Available at <http://www.ibwc.state.gov/Files/Minutes/Min270.pdf> (accessed June 25, 2008).

Minute 274, Joint Project for Improvement of the Quality of the Waters of the New River at Calexico, California-Mexicali, Baja California. 1987. April 15. Available at <http://www.ibwc.state.gov/Files/Minutes/Min274.pdf> (accessed June 25, 2008).

Minute 283, Conceptual Plan for the International Solution to the Border Sanitation Problem in San Diego, California/Tijuana, Baja California. 1990. July 2. Available at <http://ocid.nacse.org/qml/research/tfdd/toTFDDdocs/270ENG.pdf> (accessed October 20, 2007).

Minute 294, Facilities Planning Program for the Solution of Border Sanitation Problems. 1995. November 24. Available at <http://www.ibwc.state.gov/Files/Minutes/Min294.pdf> (accessed June 25, 2008).

Minute 296, Distribution of Construction, Operation, and Maintenance Costs for the International Wastewater Treatment Plant Constructed under the Agreements in Commission Minute 283 for the Solution of the Border Sanitation Problem at San Diego, California/Tijuana, Baja California. 1997. April 16. Available at <http://www.ibwc.state.gov/Files/Minutes/Min296.pdf> (accessed June 25, 2008).

Minute 298, Recommendation for Construction of Works Parallel to the City of Tijuana, B.C. Wastewater Pumping and Disposal System and Rehabilitation of the San Antonio De Los Buenos Treatment Plant. 1997. December 2. Available at <http://www.ibwc.state.gov/Files/Minutes/Min298.pdf> (accessed June 25, 2008).

Myers, Norman. 1993. *Ultimate Security: The Environmental Basis of Political Stability*. New York: W. W. Norton.

Naff, Thomas. 1994. "Conflict and water use in the Middle East." In *Water in the Arab World: Perspectives and Prognoses*, ed. Peter Rogers and Peter Lydon, 253–284. Cambridge, MA: Harvard University Press.

Naff, Thomas, and Fredrick Frey. 1985. "Water: An emerging issue in the Middle East." *Annals of the American Academy of Political and Social Science* 482 (1): 65–84.

Naff, Thomas, and Ruth Matson, eds. 1984. *Water in the Middle East: Conflict or Cooperation*. Boulder, CO: Westview Press.

Nile River Agreement (Agreement between the Republic of the Sudan and the United Arab Republic for the Full Utilization of the Nile Waters). 1959. November 8. Available at <http://ocid.nacse.org/tfdd/tfdddocs/110ENG.htm> (accessed August 22, 2008).

Nishat, Ainun, and Islam Faisal. 2000. "An assessment of the institutional mechanisms for water negotiations in the Ganges-Brahmaputra-Meghna Basin." *International Negotiation* 5 (2): 289–310.

Ohlsson, Leif. 1999. "Environment, scarcity, and conflict: A study in Malthusian concerns." PhD diss., University of Göteborg, Sweden.

Ohlsson, Leif, and Anthony Turton. 2000. "The turning of a screw: Social resource scarcity as a bottle-neck in adaptation to water scarcity." *Stockholm Water Front* 1:10–11.

Oregon State University, Transboundary Freshwater Dispute Database. n.d. *International Freshwater Treaties Database*. Available at <http://www.transboundarywaters.orst.edu/database/interfreshtreatdata.html> (accessed March 23, 2007).

Ostrom, Elinor, Joanna Burger, Christopher Field, Richard Norgaard, and David Policansky. 1999. "Revisiting the commons: Local lessons, global challenges." *Science* 284 (5412): 278–282.

Oye, Kenneth. 1986. "Explaining cooperation under anarchy: Hypotheses and strategies." In *Cooperation under Anarchy*, ed. Kenneth Oye, 1–24. Princeton, NJ: Princeton University Press.

Pachova, Nevelina, Mikiyasu Nakayama, and Libor Jansky, eds. 2008. *International Water Security: Domestic Threats and Opportunities*. Tokyo: United Nations University Press.

Pacific Institute. n.d. *Water and Conflict Chronology*. Available at <http://www.worldwater.org/chronology.html> (accessed October 30, 2007).

Parana River Agreement (Treaty between the Federal Republic of Brazil and the Republic of Paraguay concerning the Hydroelectric Utilization of the Water Resources of the Paraná River Owned in Condominium by the Two Countries, from and Including the Salto Grande de Sete Quedas or Sal to del Guaira). 1973. April 26. Available at <http://ocid.nacse.org/qml/research/tfdd/toTFDDdocs/152ENG.htm> (accessed June 5, 2008).

Protocol on Matters Pertaining to Economic Cooperation between the Republic of Turkey and the Syrian Arab Republic. 1987. *Article* 6 (July).

Putnam, Robert. 1998. "Diplomacy and domestic politics: The logic of two level games." *International Organization* 42 (3): 427–460.

Rangarajan, L. N. 1985. *The Limitation of Conflict: A Theory of Bargaining and Negotiation.* New York: St. Martin's Press.

Rawls, John. 1971. *A Theory of Justice.* Cambridge, MA: Belknap Press of Harvard University Press.

Rhine River Agreement (Convention on the Protection of the Rhine against Pollution by Chlorides). 1976. March 12. Available at <http://www.ecolex.org/en/treaties/treaties_fulltext.php?docnr=2348&language=en> (accessed March 3, 2007).

Rogers, Peter. 1993. "The value of cooperation in resolving international river disputes." *Natural Resources Forum* 17 (2): 117–131.

Rogers, Peter, and Peter Lydon, eds. 1994. *Water in the Arab World: Perspectives and Prognoses.* Cambridge, MA: Harvard University Press.

Rosegrant, Mark. 2001. "Dealing with water scarcity in the 21st century." In *The Unfinished Agenda: Perspectives on Overcoming Hunger, Poverty, and Environmental Degradation*, ed. Per Pinstrup-Andersen and Rajul Pandya-Lorch, 145–150. Washington, DC: International Food Policy Research Institute.

Salman, Salman, and Kishor Uprety. 2002. *Conflict and Cooperation on South Asia's International Rivers.* Washington, DC: World Bank.

Scheumann, Waltina, and Manuel, Schiffler, eds. 1998. *Water in the Middle East: Potential for Conflicts and Prospects for Cooperation.* Berlin: Springer.

Schwabach, Aaron. 1998. "The United Nations Convention on the Law of Non-navigational Uses of International Watercourses, customary international law, and the interests of developing upper riparians." *Texas International Law Journal* 33: 257–279.

Seyler, Daniel. 1990. "The economy." In *Paraguay: A Country Study*, ed. Dennis Hanratty and Sandra Meditz. Washington, DC: Library of Congress. Available at <http://lcweb2.loc.gov/frd/cs/pytoc.html> (accessed March 8, 2008).

Shrestha, Hari Man, and Lekh Man Singh. 1996. "The Ganges-Brahmaputra system: A Nepalese perspective in the context of regional cooperation." In *Asian International Waters: From Ganges-Brahmaputra to Mekong*, ed. Asit Biswas and Tsuyoshi Hashimoto, 81–94. Bombay: Oxford University Press.

Sigman, Hillary. 2002a. "Federalism and transboundary spillovers: Water quality in US rivers." Working paper, Department of Economics, Rutgers University.

Sigman, Hillary. 2002b. "International spillovers and water quality in rivers: Do countries free ride?" *American Economic Review* 92 (4): 1152–1159.

Starr, Joyce. 1991. "Water wars." *Foreign Policy* 82: 17–36.

Stein, Arthur. 1982. "Coordination and collaboration: Regimes in an anarchic world." *International Organization* 36 (2): 299–324.

Stein, Arthur. 1990. *Why Nations Cooperate: Circumstances and Choice in International Relations*. Ithaca, NY: Cornell University Press.

Teclaff, Ludwik. 1967. *The River Basin in History and Law*. The Hague: Martinus Nijhoff.

Tir, Jaroslav, and John Ackerman. 2009. "Politics of formalized river cooperation." *Journal of Peace Research* 46 (5): 623–640.

Toset, Hans Petter Wollebæk, Nils Petter Gleditsch, and Håvard Hegre. 2000. "Shared rivers and interstate conflict." *Political Geography* 19 (8): 971–996.

Treaty between the United States of America and Mexico Relating to the Utilization of the Waters of the Tijuana and Colorado Rivers and of the Rio Grande. 1944. November 14. Available at <http://ocid.nacse.org/qml/research/tfdd/toTFDDdocs/55ENG.htm> (accessed March 12, 2008).

Turton, Anthony, and Roland Henwood, eds. 2002. *Hydropolitics in the Developing World: A Southern African Perspective*. Pretoria: African Water Issues Research Unit.

UNESCO (United Nations Educational, Scientific, and Cultural Organization). n.d. *From Potential Conflict to Co-operation Potential*. Available at <http://www.unesco.org/water/wwap/pccp/> (accessed September 14, 2007).

United Nations Convention on the Law of the Non-navigational Uses of International Watercourses. 1997. Available at <http://untreaty.un.org/ilc/texts/instruments/english/conventions/8_3_1997.pdf> (accessed March 17, 2008).

Vener, Berrin, and Michael Campana. 2010. "Conflict and cooperation in the South Caucasus: The Kura-Araks Basin of Armenia, Azerbaijan, and Georgia." In *Water, Environmental Security and Sustainable Rural Development: Conflict and Cooperation in Central Asia*, ed. Murat Arsel and Max Spoor. London: Routledge, 144–174.

Verghese, B. G. 1996. "Towards an eastern Himalayan Rivers concord." In *Asian International Waters: From Ganges-Brahmaputra to Mekong*, ed. Asit Biswas and Tsuyoshi Hashimoto, 25–59. Bombay: Oxford University Press.

Victor, David, Kal Raustiala, and Eugene Skolnikoff. 1998. "Introduction and overview." In *The Implementation and Effectiveness of International Environmental Commitments: Theory and Practice*, ed. David Victor, Kal Raustiala, and Eugene Skolnikoff, 1–46. Cambridge, MA: MIT Press.

Waltz, Kenneth. 1979. *Theory of International Politics*. Reading, MA: Addison-Wesley.

Weinthal, Erika. 2002. *State Making and Environmental Cooperation: Linking Domestic and International Politics in Central Asia*. Cambridge, MA: MIT Press.

Westing, Arthur, ed. 1986. *Global Resources and International Conflict*. Oxford: Oxford University Press.

Wolf, Aaron. 1998. "Conflict and cooperation along international waterways." *Water Policy* 1 (2): 251–265.

Wolf, Aaron, and Jesse Hamner. 2000. "Trends in transboundary water disputes and dispute resolution." In *Water for Peace in the Middle East and Southern Africa*, 55–66. Geneva: Green Cross International.

Working Group II of the Navigating Peace Initiative. n.d. "Water conflict and cooperation: Looking over the horizon." Environmental Change and Security Program. Available at <http://www.wilsoncenter.org/index.cfm?topic_id=1413&fuseaction=topics.item&news_id=22578> (accessed June 8, 2009).

Yoffe, Shira, Aaron Wolf, and Mark Giordano. 2003. "Conflict and cooperation over international freshwater resources: Indicators of basins at risk." *Journal of the American Water Resources Association* 39 (5): 1109–1126.

Young, Oran. 1994. *International Governance: Protecting the Environment in a Stateless Society*. Ithaca, NY: Cornell University Press.

Yu, Winston. 2008. "Benefit sharing in international rivers: Findings from three case studies in the Senegal River Basin, the Columbia River Basin, and the Lesotho Highlands Water Project." Africa Region Water Resources Unit Working Paper 1. Washington, DC: World Bank.

Zawahri, Neda. 2009. "Third party mediation of international river disputes: Lessons from the Indus River." *International Negotiation* 14 (2): 281–310.

9

Is Oil Worth Fighting For? Evidence from Three Cases

Christopher J. Fettweis

At some point in the twenty-first century, the world will begin to run low on oil. Demand around the world is skyrocketing for the nonrenewable resource, far outpacing the growth of the supply, and most projections indicate that the pace will continue. While oil will not likely ever run out in the literal sense, geologists warn that in the not-so-distant future oil may well be a relatively scarce commodity. The pressure on states to secure a stable supply will increase; instability and even conflict could be the result.

Since the end of the Cold War, a number of scholars have suggested that important clues about the future of international politics can be found in petroleum geology. Scarcity might be the defining feature of the twenty-first century, spawning a series of "resource wars" as states seek to secure a steady supply of the commodities that have become vital to their security and economy. If the world contained an unlimited bounty of natural resources, the current era of great power peace might continue indefinitely; unfortunately, there may come a time when states feel the need to go to war to secure safe supplies. Since the single most vital national interest for today's industrial (and postindustrial) states is access to oil at a stable price, as oil becomes scarce, great power peace will be put to the test.

This chapter examines the litany of pessimistic predictions about future resource scarcity and competition. It evaluates the evidence from three cases of international oil interaction in an attempt to understand how states have traditionally acted regarding the most crucial natural resource. Considering the regions of the Persian Gulf, Caspian Sea, and Pacific Rim, this chapter finds that cooperation rather than conflict is the rule concerning even the most vital of national interests. Scarcity is at the heart of this analysis, yet beyond the sheer availability of the resource, this chapter touches on key economic and political factors for explaining why pessimistic predictions may be misguided. While country asymmetries seem to play an unimportant explanatory role, the sheer desire for market stability, the need to resolve the legal status of an oil-rich disputed territory, a requisite agreement on the location of pipelines and export routes, great power interests, a new leadership in a given state and domestic politics, and structural constraints are argued to be significant.

Oil, War, and International Relations Theory

Pressure on the global petroleum supply is likely to grow as this new century unfolds. No matter how much the world invests in alternative fuels, conservation or efficiency, both private and public projections foresee a steady growth in demand accompanied by a concomitant diminution of supply. Per capita energy use may hold steady or even decline across much of the industrialized world, but the projected growth in population will more than compensate. Based on the reference or mid-range projection of the U.S. Energy Information Agency (2006), even with higher prices world oil use will grow from 80 million barrels per day in 2003 to 98 million barrels per day in 2015 and 118 million by 2030. Such growth would obviously require a major increase in the current production capacity of the industry.

A number of scholars and analysts have argued that if history is any guide, the supply of oil will probably be able to keep pace with this rising demand (Manning 2000a; Odell 2004; Clarke 2007). Daniel Yergin and his colleagues at the Carnegie Energy Associates have contended that the current oil price hikes constitute nothing more than a short-term crisis—one that will soon be alleviated as market forces propel increased exploration for new supplies and alternative fuels technology, both of which will help keep prices under control.[1]

Generally speaking, however, pessimism dominates both the popular and scholarly debate about the future of oil (Orme 1997/1998; Homer-Dixon 1999; Deudney and Matthew 1999; Klare 2001; Salameh 2001; O'Hanlon 2001; Pegg 2003; Deffeyes 2003; Heinberg 2003; Campbell 2004, 2005; Roberts 2004; Simmons 2006). One scholar echoed the thoughts of many when he maintained that "it is often held that the Gulf War of 1991 was the first of a new species of 'resource wars' that will become more and more common as ecological and environmental pressures build up and affect the international political economy" (Matthews 1993, 191). In coming decades, heightened competition over dwindling resources may well lead to both international and internal conflict. "International competition for access to vital materials," argued Michael Klare (2004, 428), "is certain to become increasingly intense and conflictive."

As a general rule, the more advanced the society, the greater its dependence on fossil fuels. Due to the uneven distribution of fossil fuels around the world, energy autarky is impossible for many states, including most of the great powers. States without significant stocks are forced to rely on the goodwill of those who control the few oil-rich regions. While territorial conquest may be rare in the twenty-first century, states may make exceptions when they feel that national prosperity is at stake. It is conceivable that a situation may arise when the imperative to assure access to oil overwhelms the norm that has otherwise rendered a major war obsolete (Mueller 1989, 1995, 2004; Ray 1989; Kaysen 1990; Rosecrance 1986, 1999; Kegley 1993; Mandelbaum 1998, 2002; Jervis 2002). Perhaps competition over fossil fuels will soon put an end to such international relations optimism.

Stephen Krasner (1978) has argued that when it comes to energy resources, realpolitik best accounts for the behavior of states (see also Doran 1977). Classical realist theory contends that states in a self-help system desire autarky, because dependence on other states creates an unacceptable amount of physical and psychological insecurity (Waltz 1979, 152–158). After all, as John Mearsheimer (1994/1995, 10) has contended, since "states can never be certain about the intentions of other states," dependence creates profound vulnerabilities. The result is that a state "worries lest it become dependent on others through cooperative endeavors and exchanges of goods and services" (Waltz 1979, 106). Since oil is one of the most vital natural resources to an industrialized society, autarky in petroleum affairs should be high on the list of desires for the self-interested state. However self-defeating a war for oil may seem in the short run, there is reason to believe that in the long run, such conflict could pay dividends (Liberman 1998).

Perceptions of oil's importance may be even greater than the nontrivial empirical economic impact it has on modern society. Oil is like oxygen to industrial age states, all of whom are susceptible to irrational action when their supply is threatened. Memories of 1973 linger in the public consciousness today, when as many observers have noted, the leading democracies were gripped by turmoil and panic when their vulnerability was exposed (Doran 1977, 101; Yergin 1991, 616; Nye 1980).[2] Since financial markets are quite vulnerable to perceptions of weakness, deep fears of oil shortages may lead to damage that is beyond the ability of pure economic rationality to explain. In an essay otherwise downplaying the importance of oil as a causal factor in the Gulf War of 1991, Shibley Telhami (1992, 98) conceded that one cannot "exclude the possibility that, whether logically or not, American officials 'perceived' a need to intervene because of oil." Indeed, oil is one of those few commodities whose significance has a cognitive or perhaps even visceral dimension, which can sometimes transcend empirical cost-benefit analyses. The idea that oil is worth fighting for appears to be commonplace. Wars, after all, need not be rational.

Oil is therefore a possible candidate, if not the single most likely one, for a resource war in the coming century. The regions most at risk for hosting such a conflict would presumably be those that have both an enormous supply of recoverable fossil fuels and some degree of territorial contestation. The presence of oil assures that the external interest in the region will be high; regional weakness or disputes provide the potential for instability and power vacuums that could metaphorically suck interfering outsiders in.

One such region is the Persian Gulf, which of course contains the largest deposits of fossil fuels on earth and a variety of outstanding border disputes. Some of the major fields are controlled by two or more states, and the region is replete with states that occasionally teeter on the verge of chaos. Another obvious case is the region surrounding the Caspian Sea, whose resources, while not as extensive as those of the Gulf, are certainly sufficient to draw the attention of the great powers. Disputes regarding the legal status of the Caspian Sea, the future disposition of export pipelines, asymmetries along the pipeline routes, and the general weakness of the new states of the region make it high on the list of "hot spots" for the ecopessimists. The final potential resource war region of the twenty-first century is the Pacific Rim. Although the size of this region's reserves is not entirely clear, its strategic position near the states of the oil-thirsty Pacific states makes it an area "ripe for rivalry," to paraphrase conventional wisdom (Friedberg 1993/1994). China, Taiwan, Vietnam, Japan, and the West all have great stakes in the evolving competition over ownership of oil under the western Pacific seabed in the coming decades. No agreement over who has the right to extract and distribute the oil under the South China, East China, or Yellow seas has yet been reached.

The three regions under consideration seem to be the most likely areas for the kind of competitive behavior that ecopessimism predicts. In methodological jargon, the three cases under consideration are "least likely" to support long-term cooperation and "most likely" to provide evidence for ecopessimist, realist theory (Eckstein 1975, 127). As Gary King, Robert Keohane, and Sidney Verba (1994, 209) have noted, the strength of the inferences made about theory "depends to a considerable extent on the difficulty of the test that the theory has passed or failed." Petroleum politics would seem to present a difficult case for optimists; the strength of the inferences it may yield would be strong indeed. How have disputes and crises in these regions unfolded? Has the use (or threat) of force been an option for the states involved? Does the relative military power of the sides in disputes have any effect on the outcomes and behavior? Overall, does major war over the territory that contains energy-producing natural resources seem to be a viable possibility?

The World's Cauldron: The Energy Politics of the Persian Gulf

One of the regions where ecopessimists think that twenty-first century resource wars are most likely to erupt is the Persian Gulf, which is home to nearly two-thirds of the world's oil. The presence of petroleum was a necessary, if not sufficient, condition for both Gulf wars, and is the main reason for great power interest in an otherwise-barren, violent, and tyrannical region. The Gulf has been on the short list of vital national interests of the United States since British influence in the region waned, and assuring free and fair access to its oil has been a cause for which the US has long been prepared to fight. As the world's thirst for oil grows throughout this century, the strategic significance of the Gulf will only increase.

Former Secretary of Defense James Schlesinger was surely not alone when he argued that outside attention and internal passions as well as rivalries have turned the Middle East into "the world's cauldron." The Gulf region had become "something akin to the Balkans before 1914," he asserted, "a potential tinder box" where "any spark must immediately attract the attention of outside forces." If World War III is to come, he felt, it will probably be fought over the oil reserves of the Middle East (Schlesinger 1990, 113–115). Daniel Deudney and Richard Matthew (1999, 208), skeptical about the relationship between scarcity and war, make an exception for the Persian Gulf, where "war over oil remains a real possibility." Would the United States be willing to fight a major power in the Gulf—or perhaps more accurately, would another big power be willing to challenge the dominance of the United States in the region? It is of course possible that China or the European Union (EU) will someday reach the conclusion that the United States cannot be relied on to preserve consumer interests in the Gulf. Is Persian Gulf petroleum still, and will it remain, the "only economic interest for which the United States may have to fight" (Waltz 1981, 52)?

Even though the Gulf has been of vital strategic significance for the past fifty years, there has never been a time when a great power war to assure access to its riches seemed imminent. Over and over again, the outside powers have acted as if they felt that the imperative to avoid direct confrontation outweighed even the potential value of the Gulf's riches (Reich 1987; Breslauer 1990). If historical trends are any indication of future potential, radical changes would have to occur to break the interconsumer stability that exists regarding the Persian Gulf at the beginning of the twenty-first century. As the following sections should make clear, all regional trends indicate that much more likely short- and long-term scenarios involve cooperation among consumers (as well as between consumers and producers), rather than war between the great oil-thirsty powers. Four times in the past—in 1973, 1979, 1990, and 2003—incidents occurred that are instructive for those seeking insight into international behavior in oil-rich regions.

1973

It is probably not much of an exaggeration to suggest that the modern oil era began as the Yom Kippur War drew toward a conclusion. The Arab oil embargo that followed rudely awakened the West to the extent and depth of its dependence on the Persian Gulf. Rather than create divisions among the great importing powers, however, the Arab states helped to solidify the notion that the real geopolitical chasm in the Persian Gulf separates consumers from producers.

Common interest among consumer states led some in the United States to argue that the great powers ought to seize Arab oil, inviting these upstart small powers to suffer what they must in opposition to overwhelming power. Such contentions gained a bit of traction when Henry Kissinger, in a *Business Week* interview, refused to rule out the use of force. It was "one thing to use it in the case of dispute over price," said the secretary. "It's another where there is some actual strangulation of the industrialized world." Although Kissinger repeatedly stated afterward that he believed that military solutions were "totally inappropriate" to the current problem, a small but important debate over the utility of using military force to seize Gulf oil ensued (Kissinger

1982, 871; Collins and Mark 1975, 79). Meanwhile, contingency plans were created in the Pentagon to seize the oil fields, should the order have been given.[3]

In an essay encouraging the United States to consider the use of force as a way to resolve the crisis, Robert Tucker (1975b, 22) noted how different this crisis was being handled than any that preceded it, due to the "absence of meaningful threat of force":

> We know how the oil crisis would have been resolved until quite recently. Indeed, until quite recently it seems safe to say that it never would have arisen because of the prevailing expectation that it would have led to armed intervention. It is important to underline this point lest the utter novelty of the present situation be lost.

In 1975, Congress commissioned a feasibility study to explore the potential for a military seizure of the Gulf's oil fields. The report concluded that such an action would be both practically and strategically unwise, for the fields would likely be damaged in any such operation, and thus assuring their long-term viability would probably prove costlier than any benefit that could be gained from their possession. More important, Soviet "abstinence from armed intervention" could not be assured. "Military operations to rescue the United States (much less its key allies) from an airtight oil embargo would combine high costs with high risks. . . . Prospects would be poor, and plights of far-reaching political, economic, social, psychological, and perhaps military consequence the penalty for failure" (Collins and Mark 1975, xx). Executive branch leaders echoed such sentiments, at least in public, and the prospect of using military force to end the oil embargo died without serious debate (Collins and Mark 1975, 77–82).

Tucker was one of the few who seemed to notice that the rules by which international politics were run had changed. In his words, "Suddenly, we find ourselves in a strange universe" (Tucker 1975b, 22), where "we are confronted with the prospect of an international society in which it may no longer be possible to insure an orderly distribution of what has been termed 'the world product'" (Tucker 1975a, 55). Tucker and other realists rejected the notion that a "political and moral transformation" was sweeping over international politics—one that was revolutionizing even matters concerning "the world product."

States remained unmoved by realist logic, though, and the odds of a war over oil remained low throughout the embargo. Years later, Tucker (1980/1981, 247) observed that in light of the decline that he and many others felt that the United States was experiencing, the Soviets had proven to be oddly cautious and tentative in their actions during the crisis. As it turns out, Moscow had come to the same conclusions as Washington about the feasibility of seizing Gulf oil. Even though the Soviets had the obvious advantage of proximity and a massive imbalance in available forces in the region, they did not seem to ever have seriously considered making such a move. Their *Command Study of Iran*, written in 1941, argued that a drive to the Gulf would be unwise due to the geography of Iran and the almost certain military response of the West (Epstein 1981). Subsequent assessments reached the same conclusion (Johnson 1989, 135). Just as the West never really considered using force to seize the world product, the Soviets did not press the advantage given to them by proximity to the Gulf, despite the persistent concern of Washington. War for oil did not appear to be worth the cost.

1979

Many people inside and outside the Carter administration were convinced that the Soviet invasion of Afghanistan in 1979 was the first step in a larger strategic design aimed at more lucrative targets. The president called the invasion the "greatest threat to peace since World War II" (Dobrynin 1995, 443), and many analysts concluded that Moscow's ultimate goal was to become the arbiter of oil supplies to the West (Brzezinksi 1983, 1988; Goldman 1980; Ross 1981; Fukuyama 1981; Thompson 1982). The United States responded by reiterating its longstanding commitments to defend the Gulf, creating the Rapid Deployment Joint Task Force, which eventually evolved into the Central Command, and establishing a permanent U.S. military presence in the Gulf. A resource war seemed to be a real possibility.

Yet recently released archival materials make it clear that the Soviet attack was not the first move in a major thrust toward the Gulf but rather an effort to remove a puppet regime in its near abroad that had grown uncooperative.[4] Longtime Soviet ambassador Anatoly Dobrynin (1995, 441) claimed in his unusually frank memoir that there was no grand strategic plan to seize a foothold on the way to the Gulf, arguing rather logically that "had the Soviet leadership possessed such a plan, it would have paid far more attention to Washington's possible reaction and would have taken preemptive diplomatic measures." Instead, the Kremlin was under the impression that Mohammad Daoud Khan, who came to power in a Soviet-sponsored coup in 1973, had switched sides and was on the Central Intelligence Agency's payroll. The Soviets apparently felt that it would be a relatively simple task to remove a potential threat on their vulnerable southern flank. There was no secret plan to thrust through Iran to the Gulf.

In fact, by the time of the invasion of Afghanistan, Soviet strategists regarded an invasion of Iran to be even less appealing than they had in 1941 (Epstein 1981, 157). Dismissing for the moment the difficulties that Iranian geography would have posed for Soviet armor, by the early 1980s the Soviets could not be sure that they would be able to count on the element of surprise. The invasions of both Czechoslovakia in 1968 and Afghanistan a decade later were detected by Western intelligence months before they occurred, and preparations for a drive to the Gulf, which would have needed to be much bigger, would surely have been detected as well (Valenta 1980). The Soviets could not expect to use a lightning strike to seize Iranian oil fields and present the West with a fait accompli. Any such attack would have necessarily risked a wider war; this was apparently enough to discourage any serious contemplation in Moscow.

There is in fact no evidence to support the idea that strategic denial of Persian Gulf oil, a longtime strategic nightmare for Washington, was ever one of Moscow's central objectives. The Soviets never even attempted to interfere with the transit of Gulf oil through the vulnerable maritime choke points, although doing so would have presumably proven both fairly simple from a military point of view and potentially devastating for the West (Bennett 1987, 115–116). U.S. analysts had long determined that there was little that the West could do to counter a Soviet invasion of Iran, but in reality few seemed to be particularly concerned about the potential for such an attack (McNaugher 1985; Fukuyama 1988).

Overall, even during the height of the Cold War, the West and the Soviets never let the omnipresent crises in the Gulf region pull them too close to the brink of conflict. Neither side seemed to feel that control of its resources was worth the fight.

1991 and 2003

As unstable as the region may seem today, it is important to note the patterns of great power interaction in the Gulf that have emerged since the Soviet Union's collapse. As the two U.S.-Iraq wars make clear, the dominant response of the consumer states has become cooperative, not confrontational, where exploitation of the Gulf's resources is concerned. The great powers have not allowed the producer states to create divisions between them and instead have managed to remain more or less united. Consumer states act as if they have internalized pacific norms of interaction rather than insecurity beget by lawlessness in a self-help system.

In 1991, Saddam Hussein sought to test the new international system and in the process misjudged the reaction of nearly every great power. The Soviets did not come to Saddam's aid; the Chinese did not block coalition action; the Japanese and Europeans, who after all were more dependent on Iraqi oil than the Americans, did not restrain their senior partner; and the United States showed the willpower to reverse the conquest of Kuwait. No great power came to assist Saddam or even blocked the U.S.-led coalition from crushing his army. More than ever before, a consumer versus producer dynamic, not interconsumer rivalry, dominated the international politics of the Gulf. In other words, the Iraqi invasion demonstrated both the erosion of divisions among the great powers and the strength of post–Cold War interconsumer unity. The Gulf War was only possible because a wider war was not.

Much less harmony preceded the war in 2003, of course. Many aspects of this ongoing conflict are controversial, but one thing is not: there was never any threat that the war would bring consumer states to war against one another. Diplomatic opposition to the actions of the Bush administration was strong in many capitals and yet no industrialized power seemed willing to contemplate military support for the Iraqis. Washington did not need to fear armed opposition to the war by anyone other than Saddam's regime in Baghdad. Despite the fact that the French, Germans, Russians, and others protested the Bush administration's "preventive war" in Iraq, none has since reacted in ways that could realistically be interpreted as hostile. Disagreements among the strong powers over policy in the Gulf will no doubt continue, but military measures to oppose the United States have never been considered by any other industrialized state.

The Persian Gulf is a fully mature resource region, since behavioral rules are set and resource exploitation can proceed unhindered. A major jolt or shock would have to occur to overcome the current stability. Given the nature and extent of the shocks that the region has experienced to date—including minor wars, revolutions, ethnic conflicts, occupation, civil wars, coups, terrorism, repression, and so on—it is hard to imagine what kind of event would have the power to overcome the pacific trends. Despite the chaos generated by the never-ending Arab-Israeli crisis and the two wars with Iraq, if current trends are any guide to the future, the risk of a great power war in the Persian Gulf is exceptionally low.

While no one can say with certainty that there will never be a situation that brings outside powers to loggerheads in the Gulf, patterns from the past seem to indicate that the possibility of such an outcome is an extremely remote one indeed. A skeptic might assert that the status quo in the Gulf has been tolerated by the other great powers because some sort of geopolitical inertia prevents them from disturbing the existing order. Perhaps the costs that would be associated with any attempt to change the status quo hinder the other powers from challenging the order imposed by the United States (Wohlforth 2002). Perhaps consumer nations would be more willing to stake a claim to energy sources in regions in earlier developmental stages, where the demarcation lines have not yet been drawn, where ownership issues remain unsettled, and where no rules of the game exist. The next two sections will examine great power interaction over two such emerging energy sources, which should be exactly the kind of areas for which the industrialized states would be most likely to compete.

The Great Game Renewed: Energy Politics of the Caspian Sea

I cannot think of a time when we have had a region emerge as suddenly to become as strategically significant as the Caspian.
—Halliburton CEO Dick Cheney, 1998

The search for alternative sources of petroleum led to the center of post-Soviet Eurasia in the 1990s. Soviet inefficiency and mismanagement as well as the choice to develop Siberian reserves rather than those of the Caspian virtually removed the region from the international petroleum consciousness by the end of the Cold War. New discoveries throughout the 1990s reawakened hopes of a potential energy bonanza coming from the heart of Mackinder's Heartland.

Caspian Sea petroleum reserves are not nearly as large as those in the Gulf. Still, the region's importance may not be a matter of exactly how big the Caspian fossil fuel reserves really are, how expensive they would be to produce, or how uncompetitive they would be in world markets. As global demand grows over the next few decades, the search for alternatives to the Gulf may well compel external interest in the Caspian out of proportion to the economic value of its oil. The potential for regional instability is likely to make the major importers hesitant to put all their eggs in the Persian Gulf basket, especially in the wake of the 9/11 attacks and war in Iraq. Consumer interest in the Caspian is unlikely to wane as long as those question marks about Caspian potential and Persian Gulf stability are not removed.

Today the Caspian is on the cutting edge of international diplomacy, made "seriously sexy," in the words of one U.S. diplomat, by backroom deals, geopolitical machinations, and renewed "Great Games" (quoted in Lieven 1999/2000, 80).[5] This excitement has been accompanied by concern, pessimism, and even predictions of doom. "If we clash," one Russian analyst wrote, "it will be on the Caspian" (Bogomolov 1995, 21). Throughout the 1990s, the region was the most serious area of direct Western-Russian contention and potential conflict (Blank 2000, 66). The Caspian is after all a region with large reserves of energy-producing natural resources in a territory marked by disputes between brand-new, fragile, and weak states. In times past, during the era of realpolitik, these factors would have almost assured international competition over control of its resources. Today, however, there is reason to believe that the various internal problems of the Caspian are less likely to result in dangerous international cleavages than to unite the great powers in an effort to address them.

Early on, it became clear that analysts of Caspian interaction were separating into two camps, which have different projections of the region's future. If the realists (or the "Great Gamers," in Laurent Ruseckas's [1998] useful shorthand) are correct, the region may well be the home of increasing tension among both local and external powers as the global demand for oil grows. Liberal analysts ("oilers," according to Ruseckas [1998]) foresee a far more peaceful evolution of Caspian politics, with cooperation a far more likely outcome than conflict.

Great gamers do not hold much hope for a peaceful resolution to the various potential or simmering crises in the Caspian, since competition, not cooperation, is the essential feature of this (and perhaps every) region (Odom and Dujarric 1995; Cohen 1996; Ahrari with Beal 1996; Kleveman 2003). Historian Stephen Blank (2005a, 2006b, 2007), who is perhaps the most prolific and widely cited of the group, has been arguing since the collapse of the Soviet Union that if the West does not become more engaged in the Caspian, it risks "losing" the region to Russian neoimperial machinations. Zbigniew Brzezinski (1997, 2004) is equally pessimistic about the future of stability in the Caspian—unless, of course, the United States becomes far more active to counteract Moscow's influence. Great gamers typically contend that Russia is unlikely to accept anything but hegemonic or neoimperial dominance in the region and that the outside powers are equally unlikely to accept Russian hegemony. The "win-win" scenarios spoken of in Washington and elsewhere merely hide their true desire for complete control of the region along with its resources. "For Russia, the United States, Turkey, Iran, Pakistan, China, and Eurasia's oil and gas producing states," Blank (2000, 65) maintained, "control of those energy sources and their transportation to market means leverage, if not control, over the producer states' destinies." If this vision of the region is the correct one, and if all sides see this region as the goal in a renewed international competition, then there is indeed a danger of conflict in the long term.

Oilers, like many traditional liberals, take issue with almost all of the assumptions present in the Great Game approach. They do not assume that conflict is inevitable in the Caspian; they speak of win-win situations, of positive rather than zero-sum outcomes; they see the Caspian system as providing incentives for cooperation, not conflict; and most of all, they do not interpret states as unitary actors (Roberts 1996; Forsythe 1996; Ebel 1997; Jaffe and Manning 1998). Oilers point out that despite overwhelming military power, Russia has not been able to attain its goals by intimidating the other actors in the region. And finally, they put international oil companies at the center of their analyses of the Caspian equation. Strobe Talbott has been one of the strongest proponents of the geoeconomic approach to the Caspian. "The era of the Great Game is over," the former Deputy Secretary of State has repeatedly said. "What is required now . . . is just the opposite: for all responsible players in the Caucasus and Central Asia to be winners" (Talbott 1997). According to this school of thought, this new Great Game need not be zero-sum.

There are two reasons to believe that the oilers are correct. First, empirical evidence abounds to suggest that early pessimistic forecasts of emerging rivalries and/or violence in what Brzezinski (1997) called the "global Balkans" were inaccurate. Even the most contentious of regional issues—especially pipeline routes and the legal status of the Caspian—have not proven capable of inspiring belligerent behavior. Furthermore, a convincing argument can be made that the Caspian was at its most dangerous in the 1990s, and that as time goes on the risk of major war in the region shrinks even further. The window of opportunity for war is closing fast. And second, the great powers have thus far defined their interests in complementary, not competing, ways. No consumer nation acts as if it benefits from unrest in the Caspian. Stable governments serve the interests of all concerned, because stability maximizes those exports that are benign or positive (oil), and minimizes those that are negative (terrorism). Since September 11, cooperative trends are even more apparent.

Pipelines and Legal Status

The "early oil" production of the Caspian when the Soviet Union collapsed was quite meager compared to the region's potential, which was often referred to as "big oil." Early on, Great Gamers recognized two major issues that had the potential to pit regional states against one another and entice outside states on behalf of their allies. Both of these issues were directly related to big oil. First and foremost, pipelines had to be constructed to bring oil to market. Because the Caspian has no outlet to the oceans, there is no easy way to get its resources to international buyers. Therefore, in order for the Caspian to realize its potential, massive investment was needed to create or improve existing infrastructure, such as rigs and platforms for extraction along with pipelines for transportation to the market. The question of who would provide that investment has sometimes pitted national against corporate interests, and illustrates nicely the tension between the realist and neoliberal approaches to the region. Analyses of potential pipeline routes tended to emphasize either the routes' significance as instruments of external control over the destiny of the region, regarding profits as incidental, or the economic viability of the various routes, treating politics only as a variable of risk.[6]

The second potentially explosive issue was the undefined legal status of the Caspian Sea (Oxman 1996; Amineh 1999; Dekmejian and Simonian 2001). The heart of the dispute is whether the Caspian, which is an entirely landlocked, salty body of water, is a sea or a lake. If the Caspian is a sea, then according to international law each riparian state can claim ownership of the seabed adjacent to its coast. If it is a lake, its riches must be shared equally by all the surrounding states. Unsurprisingly, the states with large oil and gas deposits close to their shores (Azerbaijan and Kazakhstan) allege that the Caspian is a sea. The states whose coastlines hold fewer deposits (Russia, Turkmenistan, and Iran) have claimed that the Caspian is a lake and therefore its resources should be shared equally between the five states. Each side constructed an argument based on various precedents in international law, some of which date back to agreements signed by the Soviets and the Iranians in 1921.

Little would be gained by repeating the intricacies of these issues, both of which have been addressed at length elsewhere. The crucial point for the purposes of this discussion is that despite the fears of pessimists, neither of these issues has come close to sparking conflict. The states of the region, in conjunction with the energy companies, have reached a series of agreements on export routes, including the well-known pipeline from Baku to Ceyhan (BTC), which started carrying Caspian oil in mid-2006. The littoral countries have also held a series of meetings on the legal status issue, the most recent of which was in Tehran in late February 2007, and they may well be close to reaching a lasting agreement. Russia has dropped its objections to considering the Caspian to be a sea, and Iran may be close to doing the same. The absence of a well-defined legal status of the Caspian Sea prevents the maximum exploitation of the region's resources, which has created strong incentives for the riparian states to settle the outstanding issues (Mojthed-Zadeh 2000, 182). Many major agreements for exploration and production, which faced seemingly insurmountable problems only a decade ago, have been reached (Rabinowitz et al. 2004).

The most important and obvious fact about Caspian geopolitics is this: no side has ever used force, or even threatened to do so, in order to bring about its preferred outcome in either the pipeline or legal status dispute. In fact, the states of the Caspian have consistently failed to act in ways that realists expected. For instance, as William Wohlforth (2004) has noted, there has been little or no balancing behavior evident in the Caspian. Balancing can take two forms: states can build up their own capabilities, which is sometimes referred to as "internal balancing," or they can align with other states, which is known as "external balancing." Neither is occurring in the Caspian. In fact, most regional trends are in the opposite direction. Between 1992 and 1998, for example, Russia experienced "what was probably the steepest decline in peacetime military spending by any major power in history" (Wohlforth 2004, 220), and no other state has built a military with the capability to vie for control of Caspian oil.

External balancing has been absent as well. The only possible exception—the Shanghai Cooperation Organization (SCO)—was founded to settle lingering border disputes and share information in the struggle against Islamic fundamentalism.[7] September 11 and the events that followed have strengthened the determination of the SCO states to coordinate their antiterrorism policies (Collins and Wohlforth 2003). The SCO is not an organization designed to provide a balance to U.S. power; to the contrary, its interests align perfectly with those of Washington, which opposes fundamentalism with equal, if not greater, fervor. Few regional experts, even inveterate Great Gamers, feel that the SCO functions as a balancing organization (Blank 2005b, 150).

Despite the pessimistic predictions to the contrary, and as oilers expected, great power politics in the Caspian have evolved without a significant military component. The relative power of the actors has not mattered in any of the outcomes, perhaps because the utility of force is clearly minimal (Lieven 1999/2000, 73). The language that the players are using may resemble traditional realpolitik, but the issues over which they are arguing—and much more important, the tools that they are using to pursue them—are entirely diplomatic and economic. In 2002, Mearsheimer (2002) wrote that policymakers often employ liberal rhetoric to mask realist action. In the Caspian, this calculation is turned on its head: states occasionally employ realist rhetoric, but they act in manifestly liberal ways.

Moreover, no matter how serious the issue may seem from the rhetoric that the actors occasionally use, the danger of conflict over either pipelines or the legal status issue is likely to shrink further as time goes on. Each year that passes without the threat of war sets precedents for the peaceful resolution of disagreements. The BTC pipeline was constructed, for instance, without the accompanying violence once feared by many (Shaffer 2005). Over the course of the coming decade, other agreements will likely emerge on how to bring big oil to market and construction may begin on new routes. Conflict over pipelines, especially between great powers, is highly unlikely now that the BTC pipeline has set a cooperative precedent, and as time goes on it will become even less plausible. Martha Brill Olcott (2001, 9), arguably the leading U.S. expert on the region, wrote that:

> it certainly seems predictable that the level of Western interest in the region will diminish once the Caspian export routes are firmed up and the constructions of pipelines begun. . . . Once pipelines are built and production begins, the focus on the region is likely to shift to potential new areas of energy exploration. There will of course be interest in maintaining the flow of oil, but relations will move to a "maintenance" phase.

This maintenance phase is unlikely to be as contentious as the initial negotiations, which though sometimes spirited, are by no means explosive. Peaceful regional trends are likely to continue as time goes on.

Common Great Power Interests

Is the emerging regional system in the Caspian zero-sum or positive sum? Perhaps no question has more importance for the potential for conflict and the region, and certainly none more clearly separates the Great Gamers from the oilers. As of now, rather than being in conflict, the interests of the outside powers seem to overlap. First and foremost, if no pipelines are able to bring Caspian oil to market, then no one benefits. But the United States, Europe, Japan, China, and Russia also have an interest in seeing that the new states do not disintegrate into civil or ethnic conflict; that they develop healthy, stable economic and political systems, or at least avoid implosion; that they do not engage in the proliferation of drugs or weapons of mass destruction; and that Islamic fundamentalists do not find a home anywhere. Far from fomenting problems, in the majority of cases, external influences presently have had a pacifying effect. Like international war, ethnopolitical conflict raises the risks as well as the costs involved with exporting the region's fossil fuels, thereby lowering the potential for big oil to truly be big. The incentive structure for outside powers is therefore tilted toward peace. Currently, it does not appear as if the great powers are inciting instability in the Caspian, nor are they prepared to become involved if and when fighting does break out in anything but a peacemaking role. Even Blank (2000, 72) admitted that by the decade's end, peace operations had replaced gunboat diplomacy as "the main military instrument for the establishment of great power spheres of influence." Another prominent Great Gamer, Ariel Cohen (2005) of the Heritage Foundation, wrote that "the potential for Russia and the U.S. to pursue a parallel foreign policy in the region—one based on interests, not emotions—is greater than many think." The great powers are willing to allow others to provide the stability that benefits everyone, since the ultimate outcome is no longer seen as zero-sum.

International relations in the Caspian are unfolding like a legal dispute, not unlike the one over the status of the sea itself. The parties presented their positions, and even though some interests came into conflict, eventually the court in which the disputes are settled has functioned as a stateroom of sorts, not a battlefield. Each side has a stake in the maintenance of order in the system, like citizen stakeholders in domestic society. The balance of military power, especially between the great powers, hardly factors into the dispute at all. Despite the fact that billions (perhaps even trillions) of dollars and tremendous international influence is at stake in the final outcome of this Caspian drama, the great powers do not seem to find it worthwhile to risk war. Oft-repeated predictions of doom are not coming true, and the likelihood of major war in the region—which was never high—is shrinking ever further as the months pass. Since great power cooperation does not make for exciting, front-page material, news from the Caspian will probably remain rather dull for some time to come.

Realpolitik's Last Stand? Energy Politics in the Pacific Rim

One of the truly significant moments in the history of international politics took place sometime in late 1993. The exact date it occurred is unclear, since at the time no one seemed to notice. There were no headlines, no news coverage, no analysis from CNN pundits, and not even hyperbolic warnings from Congress. Looking back, it seems remarkable that no one in 1993 took note of the moment that the People's Republic of China (PRC) became a net importer of oil (Salameh 1995, 141). Few events were to have as much lasting significance for economics, politics, and national security affairs, as the transition to oil importer status was an early symptom of the rapid growth that the Chinese economy was to experience over the next decade (and counting) (Calder 1996; Manning 2000b). The implications for international politics are still unclear.

China's meteoric rise has not gone unnoticed in economic and military circles around the world. More than a few observers warn that this growing economic prowess and its concomitant rising demand for oil may put the PRC on a collision course with the rest of the world in general, and the United States in particular (Friedberg 2005). One cannot help but think, for example, that if the national energy company of Australia had attempted to buy a controlling interest in a relatively minor U.S. oil corporation in summer 2005, the few that actually took notice would have had to search the back pages of the business sections in newspapers for details. But when China's national energy company attempted to take over Unocal, it became front-page news with supposed national security implications that were the talk of scholars, analysts, and members of Congress. The reaction to the proposed purchase seemed to surprise Beijing, which soon backed off from its efforts (Klare 2006, 183).

A trio of island chains—the Paracels, Diaoyus, and the loosely connected Spratlys—lie at the heart of international attention in the area.[8] All three are more accurately thought of as collections of rocky outcroppings or atolls that often disappear at high tide, since none contains a single island capable of sustaining human habitation. Nonetheless, control of these islets has become equivalent to a de facto control of the sea and the resources that may lie beneath it. Unsurprisingly, competing international claims to sovereignty over these islands and the surrounding seas did not emerge until estimates of the oil reserves under the seabed were first released in the late 1960s (Harrison 2002). South Vietnam was one of the first to react, awarding a contract for oil exploration to foreign oil companies in July 1973 (Catley and Keliat 1997). The PRC quickly produced ancient Han Dynasty references to the islands, supposedly proving that all the chains had been Chinese for two millennia (Jinming and Dexia 2003). This precipitated something of a scramble for control of the islets, with the various claimants solidifying their positions by sporadically building a variety of semipermanent structures, like the meteorologic station that China built on Fiery Cross Reef in 1987 (Dobson and Fravel 1997, 259). Beijing in particular, as one observer noted, "appears fully conscious of the maxim that possession is nine-tenths of the law" (Leifer 1995, 48). The potential for conflict over the islands seemed to increase dramatically in February 1992, when China passed a law asserting sovereignty over the Spratlys, Paracels, Diaoyus, and all other islands in the South and East China seas, and warned that it would be prepared to defend them (Salameh 1995). The region has been operating under a fairly precarious peace ever since.

Traditional realist analysis would suggest that the growing economic power of China will inevitably be turned into military power, which when coupled with its new thirst for oil, will lead to expansion and perhaps conflict with the United States (Layne 2004). According to this litany of pessimistic projections, China will seek to maximize its power and influence throughout the next century, and will do so by military means whenever necessary (Christensen 1999, 2001; Betts 1993/1994; Garver 1997; Waldron 1997; Bernstein and Munro 1997; Gertz 2000; Carpenter 2006). Such assessments obviously do not place much faith in the potential for a fundamental transformation of the international system. If the world operates according to the basic assumptions of realism, the rise of China will indeed be accompanied by balancing, suspicion, security dilemmas, and instability. The oil supplies of the region would just add to the region's problems, inspiring self-interested littoral states to vie for their control. Christopher Layne (2004, 74) holds that a U.S.-China rivalry is "highly likely" while Mearsheimer and Brzezinski (2005, 47) argue that China "cannot rise peacefully." If the realist vision of the future is correct, China probably cannot do so.

Many longtime China watchers have asserted that the PRC is hyper-realist, the "high church of realpolitik," motivated primarily by self-interest and beholden to power politics (Christensen 1996, 37). Alastair Johnston's sweeping examination of Chinese history (1995) suggests that a "hard realpolitik paradigm" is embedded in China's strategic culture, which has manifested itself in Chinese foreign policy from the Han Dynasty two millennia ago right up through Mao Tse-tung and his successors. Bob Catley and Makmur Keliat (1997, 207) contend that rather than pursue twenty-first century political reforms, the Chinese leadership has decided to "reinvent the nineteenth century, Bismarckian, militaristic, nationalist state," which has been the necessary and sufficient cause of international gridlock in the nearby seas. Chinese intransigence over the various disputed islands is certainly an important data point in the assessment of this hard realpolitik paradigm. "One thing is certain," this pessimistic argument goes. "China will be as robust as the United States in defending its access to oil supplies. Furthermore, China may not shy away from the use of force to defend its rights of access" (Salameh 1995, 141). The growing PRC may not remain satisfied with the status quo in its nearby oil-rich seas, especially now that it has lost autarky in energy affairs.

Many issues crucial to energy politics in the Pacific Rim remain unsettled, from basic questions about the size of undersea stockpiles to more complicated and controversial ownership issues. Yet there are two major reasons to believe that the potential for such a clash, even in this early, chaotic stage, is fairly small. First, a new generation of Chinese leadership seems to have changed Beijing's approach to foreign policy. These leaders, the first who were not contemporaries of Chairman Mao, and have no memory of the ideological struggles of the 1950s and 1960s, have embarked on a regional "charm offensive" and certainly act as if they believe that major war has become obsolete. Second, when forced to choose between nationalism and the performance of their economy to justify the legitimacy of their regime, this new Chinese leadership has consistently chosen the latter. Finally, a series of structural constraints exist that seem to support the contention that a norm against warfare exists in the Pacific Rim.

New Leaders, New Priorities

As it turns out, even the high church of realpolitik is not a unitary actor. A variety of forces battle inside Beijing for control of Chinese foreign policy, and its actions are reflective of the outcome of that struggle.[9] Although clearly there are still many who fear China's hidden and perhaps aggressive motives, over the last five years a number of well-respected China watchers have noted that the trends in Chinese domestic politics may be stacking the deck in favor of cooperative international outcomes.

Some scholars have suggested that increases in Chinese nationalism (and belligerence) throughout the 1990s may well have been largely the result of the high degree of influence that the People's Liberation Army (PLA) had on decision making (Shambaugh 1999/2000). John Garver (1996) has argued that in the beginning of the decade, military officers were quite influential in determining the direction of Chinese policy toward the South China Sea. There is evidence, for instance, that the Chinese occupation of Mischief Reef in 1995 occurred without the authorization of the civilian leadership in Beijing (Fravel 2001, 8; Zha and Valencia 2001). The Ministry of Foreign Affairs has clashed repeatedly with the PLA over foreign policy matters, in particular concerning policy toward the Spratly Islands (Fravel 2001).

Since about 1997, however, when Beijing instituted a series of major military reforms, the influence of the PLA seems to have waned (Storey 1999). If indeed belligerent rhetoric and actions from the PRC is the result of the PLA's undue influence, and that influence is decreasing, then China can be expected to act like a status quo power in the various island disputes on its periphery (Johnston 2003). Either way, the infighting between the PLA and the Ministry of Foreign Affairs casts doubt on the notion that Chinese policy is the result of a unitary actor/hard realpolitik paradigm native to its strategic culture.

The diminution of PLA influence on policymaking is indicative of a broader generational change that appears to be happening inside Beijing. A number of China experts have begun to claim that the current leadership of the PRC has little in common with the founding members of the Communist Party, and is far less dogmatic in its approach to both economics and politics (Christensen and Glosny 2004). While it may be a bit premature to suggest that there is a Chinese Gorbachev ready to bring political freedom to his people, at the very least Beijing has altered the way it treats its neighbors. China's much-discussed "charm offensive" has won it many friends in East Asia, and has helped solidify many of the complex economic ties that cement stability across the region and avoid the regional tensions that realists have expected to see in response to its rapid economic growth (Kurlantzick 2007; Medeiros and Fravel 2003; Shambaugh 2004/2005). Beijing has been reluctant to use its military superiority to threaten or bully its neighbors into capitulation. Instead it has consistently chosen a more cooperative, positive-sum path.

Legitimacy

Many China watchers fear that the regime began to experience a crisis of legitimacy in the early 1990s, when inefficiency and Tiananmen Square effectively discredited the four-decades-old Communist experiment (Downs and Saunders 1998/1999, 118). In the international society of the late twentieth century, governments did not find it easy to rule absent a clear mandate from the masses, which the rulers in Beijing did not possess. Throughout the 1990s, the Communist Party began to rely on two powerful new sources, the "twin pillars," to legitimize its undemocratic rule: nationalism and economic growth.[10] At times, these two pillars pulled the regime in different directions. On the one hand, if the Communist oligarchy appeals to Chinese nationalism to legitimize its rule, then it can be expected to defend its interests quite aggressively. If, on the other hand, it decides that continued economic growth is the key to remaining in power (so-called performance legitimacy), then Chinese behavior will likely be moderate, integrative, and cooperative in order to keep the economy humming. In a real sense, the future of stability in the Pacific Rim hinges on the balance struck by the Chinese leadership and the path down which it leads the country.

While finding a steady source of oil will clearly be vital to Chinese leaders, the desire for petroleum will not necessarily lead them down the warpath. Despite the fact that some longtime China observers have argued that "seizing the initiative is embedded in doctrine as a preferred course of action," a major conflict in the Pacific Rim would likely end the Chinese economic miracle, crippling the leaders' hopes of escape from the Third World (Whiting 2001, 105). In the few times since the end of the Cold War when Beijing's twin pillars of legitimacy have seemed to come into conflict, economic considerations have outweighed the imperatives of nationalism. When forced to choose, Chinese leaders have repeatedly pursued economic development at the expense of nationalist goals (Downs and Saunders 1998/1999, 117). If this trend proves to have staying power, then war in the Pacific Rim will remain a low-probability event for the foreseeable future.

Two conflicts over the Diaoyu Islands have been particularly instructive in terms of the relative weight that Chinese leaders put on economic stability. No one seemed to have much cared who owned these barren islets until 1969, when oil was discovered under the East China Sea. The estimates of between ten to one hundred billion barrels encouraged China, Japan, and Taiwan to assert that the chain had long been part of their territory (Downs and Saunders 1998/1999, 126).[11] In 1990 and again in 1996, right-wing Japanese youth groups began to construct lighthouses on three of the Diaoyus, and Chinese nationalists on both sides of the Taiwan Strait responded with outrage. Chinese leaders initially fueled the flames of nationalism while simultaneously pursuing expanded economic relations with Tokyo (Downs and Saunders 1998/1999, 139–140). As the crisis escalated and these two goals came into conflict, however, Beijing not only decided that its economic relations outweighed its nationalist goals but also actively tried to restrain the more bellicose elements of its society, placing controls on email servers in universities and cities where anti-Japanese sentiment was growing (Guoxing 1998, 108). The formerly Communist government of the PRC now ties its right to rule on the economic performance of the country, hyperrealism notwithstanding.

Structural Constraints

Walter Lippmann once compared the United States and the Soviet Union to a whale and an elephant: while each dominates its own arena, their interests are so diverse that they need not come into conflict (Baer 1994). A similar situation exists today in the oil-rich western Pacific. The Chinese elephant and the U.S. whale may never fully trust each other, but a war between them does not seem likely, since the region contains no logical flash points.

As Robert Ross (1999) argued in one of the most insightful and persuasive pieces written on the international politics of the Pacific Rim, geography stacks the deck against war. Even if hostility does grow between the two as China rises, the rivalry would pit a state that clearly acts as a sea power (at least in this region) against one whose power lies on the land. The PRC is a land power, unlikely to be able to project its power offshore; the United States is a maritime power in Southeast Asia, with no strategic interest in competing for influence on the mainland. Logical battlegrounds where the interests of the two overlap are few and far between. The only way a clash between the two regional powers would occur is if one party intrudes on the sphere of the other—an action that neither has shown an eagerness to do. While China may be significantly stronger than it was twenty years ago, its military does not pose much of a threat to the United States on the high seas, and Beijing displays little urgency to change that calculation (Shambaugh 2005).

The second structural constraint involves the utility of war to seize offshore oil rigs. Some scholars have long held that war for the control of oil reserves would be a self-defeating proposition, since the cost involved in replacing the inevitable damage would outweigh the benefits that could be gained by conquest (Jaffe and Manning 1998). This seems to be especially true for offshore infrastructure, which is simultaneously more expensive and more vulnerable. Although the short-term economic costs of a war to gain control of the territory that contains fossil fuels can in theory be outweighed by the long-term benefits of resource exploitation, this general rule does not apply to sea-based petroleum fields like those in the South and East China seas. Any infrastructure that would be constructed to extract oil and gas from the region would be extremely fragile and difficult to secure, since the sea offers no natural protection. Oil rigs make easy targets. Disrupting their productivity would not be outside the capabilities of even small, relatively weak states.

In order for any energy company to be interested in developing the resources of this region, jurisdictional issues must be settled. As long as higher risks mean higher costs, the perception of instability will remain the single greatest factor driving potential investors away from the Pacific Rim's energy resources (Yergin, Elkof, and Edwards 1998, 47–48). No state will be able to benefit from the oil and gas fields of the region until the ownership issues are settled. As the surprisingly few rational choice analyses of the region have concluded, economic factors play a major role in encouraging China and the other regional actors to seek peaceful, negotiated resolutions of the outstanding issues (Wu and Bueno de Mesquita 1994). In order to maintain its high growth rates, Beijing may soon need to cut a deal regarding its nearby seas.

International precedents for offshore oil exploitation certainly suggest that Pacific Rim issues could be settled peacefully. In fact, there has never been a war over offshore oil deposits, even though there are several rather significant fields that have been discovered in the past few decades, from the North Sea to the Gulf of Mexico. In all cases, agreements have been reached to develop the oil and gas fields without conflict. Peaceful precedents do not guarantee peaceful futures, of course—Norway and the United Kingdom are obviously quite different from China and Taiwan—but it is worth noting that when vast offshore hydrocarbon fields have been discovered before, despite the energy autarky and billions of dollars at stake, lasting agreements have emerged that benefit all parties.

A Chinese challenge to the status quo in the Pacific would entail an enormous risk for a questionable reward, which is a calculation that Beijing seems to have made. In the East China Sea, China and Japan have taken active steps to begin developing what has been called a conflict-avoidance regime to address their many overlapping claims, related to fishing rights, scientific research notifications, and ultimately military and intelligence activities (Valencia and Amae 2003). Since the September 11 terrorist attacks, the Chinese relationships with both the United States and the Association of Southeast Asian Nations have improved dramatically, leading to the signing of a "code of conduct" for parties in the various South China Sea disputes (Song 2003; Thao

2003). Regional trends suggest that the risk of war seems to be decreasing as time goes by. Johnston (2003) is not alone among major China scholars in believing that fears of a rising dragon are misplaced, and indeed China exhibits all the signs of being a status quo power. The idea that oil is not worth fighting for may have taken hold in the Pacific Rim.

The first decade and a half of post–Cold War international relations in the Pacific Rim have not unfolded as early pessimist forecasts predicted. In fact, the states of the region have acted almost as if they were unaware of the inevitability of rivalry. No alarms seem to have been rung in response to the growth of China, and even the high church has not followed the basic expectations of realpolitik. The post–Cold War period has been marked by a notable (and to realists, puzzling) lack of balancing behavior in East Asia (Kang 2003). Today, both the evidence and theoretical logic support the belief that a major war to assert control over the potentially vast petroleum deposits in the South and East China seas, despite lingering disputes over their ownership and rapidly increasing regional demand, is not likely in the indefinite future. If indeed the use of force to assure access to oil is not a realistic option for even the high church of realpolitik in a region at an early, unstable stage of petroleum politics, then can it be an option anywhere?

The Stages of Petroleum Politics

The Persian Gulf has been central to petroleum politics throughout the entire oil era; the Caspian appeared on the energy radar screen only after the fall of the Soviet Union; much of the story of Pacific Rim oil has yet to be written. Taken together, the cases present a full story of the international relations of petroleum and lend a great deal of insight into what the future can be expected to bring. The politics of oil-rich regions seem to go through similar stages of development. None of these stages are particularly dangerous; in fact, each is more stable than the last.

The most dangerous stage for any weak, petroleum-rich region is the first, which extends from the discovery of the existence of the fuels through the initial squabbles over who will own the rights to their development. Scholars and strategists will likely be most vociferous in their debates over the potential for problems during this first stage, and for good reason, since it is the most unstable period, the time when war is most likely as actors jockey for influence over the exploitation to come. The economic and strategic stakes can hardly be higher than establishing a precedent over the control of petroleum-rich territory. The Pacific Rim clearly falls into this category. Not only are reliable estimates of the total reserves not yet available but basic questions about ownership and distribution rights are far from settled as well. Petroleum production in the region is at an inchoate stage, and many uncertainties remain over exactly who will be the primary beneficiaries of the seabed's resources. Additional factors such as domestic politics and structural constraints should also be taken under consideration in this region.

Once littoral states and interested outside actors reach the initial (if fragile) development agreements in a petroleum-rich, weak region, the second stage of the life cycle begins. In this stage the oil starts to reach the market in serious amounts, often following a round of initial deal making between the interested parties. Behavioral norms and expectations begin to settle over the region, and measurable trends regarding cooperation and/or conflict emerge as parties on all sides realize who the likely winners are of the early competition. Desperate measures could be undertaken to avoid unfavorable, semipermanent outcomes that appear to be more likely with every passing year. The Caspian region is currently in this second stage, and as such its development lies neatly between that of the other two cases under consideration. While the issues of sovereignty over the undersea resources are not fully settled, they are close. A variety of agreements have been reached over the last decade to bring Caspian oil to market, based on realistic assessments of the size of its deposits rather than the wild conjecture that appeared shortly after independence. While the region has not reached what could be considered fully mature relations, since many of its major issues are not settled to the satisfaction of all parties, its level of development ought to offer clear insight into what can be expected from the middle stages of oil politics.

If no actor seems poised to make a desperate move during this period, the development of the oil-producing region moves into its final stage—when precedent has been set for the politics surrounding resource exploitation. Resource development on a large scale begins in the third stage of this political life cycle, which also should be accompanied by the smallest danger of major war. This stage starts once the sovereignty issues are all but settled, and the interests of both consumers and producers align in a desire to keep the pumps operational. Mutually satisfactory rules of the game have been established; a miniature international society determines the bounds of interaction for consumers and producers alike. The Persian Gulf clearly fits into this category. It has progressed through the first two stages, and has been the epicenter of oil development for almost as long as industrial society has been petroleum dependent, virtually synonymous with petroleum politics since the dawn of the oil age. Although it may seem quite chaotic at times, the oil politics of the Gulf are actually extremely stable, and oil flows uninterrupted by the region's occasional turmoil. Even if small wars continue, the odds of a major war breaking out in the Gulf, the richest territory in the world, is vanishingly small. The international political development of the Gulf has reached nearly full maturity, and can perhaps serve as a harbinger of things to come for the other two.

If the cases above are any indication, no stage of this life cycle carries much risk of major war to control resources. The most obvious observation that emerges from the study of petropolitics is that at no time have great powers come close to loggerheads over the control of these vital regions. No country has ever actively prepared to conquer these weak regions, nor has any felt it necessary to prepare to defend them. Consumer cooperation rather than conflict is the rule.

Conclusion

Why has great power behavior failed to live up to realist expectations? While it is hard to argue that democracy has helped confound the various ecopessimistic projections, since not all interested parties are democracies, other rationalist explanations for stability cannot be entirely ruled out. Perhaps it is the fear of escalation toward a nuclear holocaust that has kept the great powers from fighting over oil. Perhaps liberal internationalists are correct, and complex interdependence should be given primary credit. Whatever the initial cause, the idea that war would be a viable option to control the most valuable regions in the world does not seem to have occurred to the great consumer nations. As time goes on, it becomes more and more unlikely that it ever will.

Resources have always been a great motivator for war. The most valuable regions—those worthy of contestation—have always been the richest. Today, that calculation seems to have changed, even regarding the most vulnerable, valuable regions on earth in a period of almost certain increasing scarcity. Reaping the economic and financial benefits of exploiting oil resources requires interstate coordination (over unresolved territorial matters or export routes, for example). This has become a great incentive for international cooperation. Other political/structural factors have also played a role in diminishing likely violent conflict over oil. It seems as if the states of the industrialized world have reached the conclusion that oil is not worth fighting one another for. Perhaps, for the first time in history, nothing is.

Notes

1. Daniel Yergin, "It's not the end of the oil age," *The Washington Post*, July 31, 2005, B7.

2. For an in-depth review of the actions taken by states in response to the oil shocks, see Ikenberry 1988.

3. A recently released British memo at the time noted the British concerns about the possibility of armed U.S. intervention. Defense Secretary Schlesinger apparently wrote to his British counterpart that the United States would not tolerate threats from "under-developed, under-populated" countries and that it was "no longer obvious" that the United States could not use force to resolve the standoff. Yet military planners prepare for many contingencies. The extent to which the United States seriously contemplated the use of force in this instance awaits declassification of the relevant U.S. documents. See Glenn Frankel, "U.S. mulled seizing oil fields in '73: British memo cites notion of sending airborne to Mideast," *The Washington Post*, January 1, 2004, A1.

4. This conclusion follows an analysis of the documents relating to the war in Afghanistan that have been released by Moscow since the end of the Cold War. Those relevant to the beginning of the war and its motivations can be found on the Cold War International History Project of the Woodrow Wilson Center's Web site available at <http://wilsoncenter.org/index.cfm?fuseaction=topics.home&topic_id=1409> (accessed November 16, 2009). The discussions and documents relating to Afghanistan deal almost exclusively with internal Afghani problems, and make neither overt nor veiled references to next steps. While it is possible that such information and planning has never been released by Moscow, the burden of proof would seem to lie with those who claim that such a smoking gun exists, but has yet to be found.

5. The booming literature on this region refers to it in a variety of ways—Central Asia, the Caucasus, the Transcaspian, the Soviet South, and so forth. The area relevant to this discussion is what is emerging as a new geographic designation, simply "the Caspian," an area "defined less by demography or geographic proximity than by geology" (Ruseckas 1998, 2).

6. The literature on potential Caspian pipelines is quite voluminous. For some of the best realist analyses, see Cornell 1999; Blank 1994, 2000; Cohen 1996; Starr 1997. For good economic analyses of the Caspian pipeline situation, see Ebel and Menon 2000; Jaffe and Manning 1998; Roberts 1996; Ebel 1997; Kober 2000.

7. In 1996, Russia, China, Kyrgyzstan, Kazakhstan, Tajikistan, and later Uzbekistan established something akin to a "Three Emperor's League" for the twenty-first century, substituting political Islam for the liberalism that threatened the monarchies of the nineteenth century; see Misra 2001.

8. The Japanese refer to the Diaoyus islands as the Senkakus.

9. For example, the January 2007 antisatellite weapon test was apparently done without the knowledge of the Foreign Ministry; see Mulvenon 2007. For earlier scholarship that casts doubt on the utility of the unitary actor assumption in the PRC, see Lewis, Di, and Litai 1991; Garver 1992.

10. For an interesting discussion of the various forms of legitimacy, see Gilley 2006.

11. See also Nicholas D. Kristof, "An Asian mini-tempest over mini-island group," *The New York Times*, September 16, 1996, A8.

References

Ahrari, Mohammed E., with James Beal. 1996. "The new Great Game in Muslim Central Asia." Institute for National Strategic Studies McNair Paper 47. Washington, DC: National Defense University.

Amineh, Mehdi P. 1999. *Towards the Control of Oil Resources in the Caspian Region*. New York: St. Martin's Press.

Baer, George W. 1994. *One Hundred Years of Sea Power: The U.S. Navy, 1890–1990*. Palo Alto, CA: Stanford University Press.

Bennett, Alexander J. 1987. "The Soviet Union." In *The Powers in the Middle East: The Ultimate Strategic Arena*, ed. Bernard Reich, 101–132. New York: Praeger.

Bernstein, Richard, and Ross H. Munro. 1997. *The Coming Conflict with China*. New York: Knopf.

Betts, Richard K. 1993/1994. "Wealth, power, and instability." *International Security* 18 (3): 34–77.

Blank, Stephen J. 1994. *Energy and Security in Transcaucasia*. Strategic Studies Institute report. September. Carlisle, PA: U.S. Army War College.

Blank, Stephen, J. 2000. "American grand strategy and the Transcaspian region." *World Affairs* 163 (2): 65–79.

Blank, Stephen J. 2005a. *After Two Wars: Reflections on the American Strategic Revolution in Central Asia*. Strategic Studies Institute. July. Carlisle, PA: U.S. Army War College.

Blank, Stephen J. 2005b. "China in central Asia: The hegemon in waiting?" In *Eurasia in Balance: The U.S. and the Regional Power Shift*, ed. Ariel Cohen, 149–182. Burlington, VT: Ashgate.

Blank, Stephen J. 2007. *U.S. Interests in Central Asia and the Challenges to Them*. Strategic Studies Institute. March. Carlisle, PA: U.S. Army War College.

Bogomolov, Pavel. 1995. "If we clash, it'll be on the Caspian." *Current Digest of the Post-Soviet Press* 47 (21): 21–22.

Breslauer, George W. 1990. "On collaborative competition." In *Soviet Strategy in the Middle East*, ed. George W. Breslauer, 3–22. Boston: Unwin Hyman.

Brzezinksi, Zbigniew. 1983. *Power and Principle: Memoirs of the National Security Advisor, 1977–1981.* New York: Farrar, Straus, and Giroux.

Brzezinksi, Zbigniew. 1988. "After the Carter doctrine: Geostrategic stakes and turbulent crosscurrents in the Gulf." In *Crosscurrents in the Gulf: Arab, Regional, and Global Interests*, ed. Richard H. Sindelar and John E. Peterson, 1–9. London: Routledge.

Brzezinksi, Zbigniew. 1997. *The Grand Chessboard: American Primacy and Its Geostrategic Imperatives.* New York: Basic Books.

Brzezinski, Zbigniew. 2004. *The Choice: Global Domination or Global Leadership.* New York: Basic Books.

Calder, Kent E. 1996. "Asia's empty gas tank." *Foreign Affairs* 75 (2): 55–69.

Campbell, Colin J. 2004. *The Coming Oil Crisis.* New York: Multi-Science Publishing.

Campbell, Colin J. 2005. *Oil Crisis.* New York: Multi-Science Publishing.

Carpenter, Ted G. 2006. *America's Coming War with China: Collision Course over Taiwan.* New York: Palgrave Macmillan.

Catley, Bob, and Makmur Keliat. 1997. *Spratlys: The Dispute in the South China Sea.* Brookfield, VT: Ashgate.

Christensen, Thomas J. 1996. "Chinese realpolitik." *Foreign Affairs* 75 (5): 37–52.

Christensen, Thomas J. 1999. "China, the U.S.-Japan alliance, and the security dilemma in East Asia." *International Security* 23 (4): 49–80.

Christensen, Thomas J. 2001. "Posing problems without catching up: China's rise and challenges for U.S. security policy." *International Security* 25 (4): 5–40.

Christensen, Thomas J., and Michael A. Glosny. 2004. "Sources of stability in U.S.-China security relations." In *Strategic Asia 2003–04: Fragility and Crisis*, ed. Richard J. Ellings and Aaron L. Friedberg, 53–80. Washington, DC: National Bureau of Asian Research.

Clarke, Duncan. 2007. *The Battle for Barrels: Peak Oil Myths and World Oil Futures.* New York: Profile Books.

Cohen, Ariel. 1996. "The new 'Great Game': Oil politics in the Caucasus and central Asia." *Backgrounder* 1065.

Cohen, Ariel. 2005. "Competition over Eurasia: Are the U.S. and Russia on a Collision Course?" Heritage Foundation Lecture Series 901. Washington, DC: Heritage Foundation. Available at <http://www.heritage.org/Research/Reports/1996/01/BG1065nbsp-The-New-Great-Game> (accessed November 16, 2009).

Collins, James M., and Clyde R. Mark. 1975. *Oil Fields as Military Objectives: A Feasibility Study.* Washington, DC: Government Printing Office.

Collins, Kathleen A., and William C. Wohlforth. 2003. "Central Asia: Defying 'Great Game' expectations." In *Strategic Asia 2003–04: Fragility and Crisis*, ed. Richard J. Ellings and Aaron L. Friedberg, 291–318. Washington, DC: National Bureau of Asian Research.

Cornell, Svante. 1999. "Geopolitics and strategic alignments in the Caucasus and central Asia." *Journal of International Affairs* 4 (2): 100–125.

Deffeyes, Kenneth S. 2003. *Hubbert's Peak: The Impending World Oil Shortage*. Princeton, NJ: Princeton University Press.

Dekmejian, Hrair R., and Hovann H. Simonian. 2001. *Troubled Waters: The Geopolitics of the Caspian Region*. New York: I. B. Tauris.

Deudney, Daniel H., and Richard A. Matthew, eds. 1999. *Contested Grounds: Security and Conflict in the New Environmental Politics*. Albany: State University of New York Press.

Dobrynin, Anatoly. 1995. *In Confidence*. New York: Random House.

Dobson, William J., and Taylor M. Fravel. 1997. "Red herring hegemon: China in the South China Sea." *Current History* 96 (611): 258–263.

Doran, Charles F. 1977. *Myth, Oil, and Politics: Introduction to the Political Economy of Petroleum*. New York: Free Press.

Downs, Erica Strecker, and Phillip C. Saunders. 1998/1999. "Legitimacy and the limits of nationalism: China and the Diaoyu Islands." *International Security* 23 (3): 114–146.

Ebel, Robert. 1997. *Energy Choices in the Near Abroad: The Haves and the Have-nots Face the Future*. Washington, DC: Center for Strategic and International Studies.

Ebel, Robert, and Rajan Menon, eds. 2000. *Energy and Conflict in Central Asia and the Caucasus*. New York: Rowman and Littlefield.

Eckstein, Harry. 1975. "Case study and theory in political science." In *Handbook of Political Science*, ed. Fred I. Greenberg and Nelson W. Polsby, 1: 94–137. Reading, MA: Addison-Wesley.

Energy Information Agency. 2006. *International Energy Outlook 2006*. June. Washington, DC: Department of Energy.

Epstein, Joshua M. 1981. "Soviet vulnerabilities in Iran and the RDF deterrent." *International Security* 6 (2): 126–158.

Forsythe, Rosemarie. 1996. "The politics of oil in the Caucasus and Central Asia: Prospects for oil exploitation and export in the Caspian Basin." Adelphi Paper 300. London: International Institute for Strategic Studies.

Fravel, Taylor, M. 2001. "China in the South China Sea: Facts in search of a theory." Paper presented at the Annual Conference of the International Studies Association, Chicago, February 20–24.

Friedberg, Aaron L. 1993/1994. "Ripe for rivalry: Prospects for peace in a multipolar Asia." *International Security* 18 (3): 5–33.

Friedberg, Aaron L. 2005. "The future of U.S.-China relations: Is conflict inevitable?" *International Security* 30 (2): 7–45.

Fukuyama, Francis. 1981. "The Soviet threat to the Persian Gulf." *Rand Papers* 6596. Santa Monica: RAND Corporation.

Fukuyama, Francis. 1988. "Soviet military power in the Middle East; or, whatever became of power projection?" In *The Soviet-American Competition in the Middle East*, ed. Steven L. Spiegel, Mark A. Heller, and Jacob Goldberg, 159–183. Lexington, MA: D. C. Heath and Co.

Garver, John W. 1992. "China's push through the South China Sea: The interaction of bureaucratic and national interests." *China Quarterly* 132 (December): 999–1028.

Garver, John W. 1996. "The PLA as an interest group in Chinese foreign policy." In *Chinese Military Modernization*, ed. Dennison C. Lane, Mark Weisenbloom, and Dimon Liu, 246–281. London: Kegan Paul.

Garver, John W. 1997. "*Face Off: China, the United States, and Taiwan's Democratization*. Seattle: University of Washington Press.

Gertz, Bill. 2000. *The China Threat: How the People's Republic Targets America*. New York: Regnery Publishing, Inc.

Gilley, Bruce. 2006. "The meaning and measure of state legitimacy: Results for 72 countries." *European Journal of Political Research* 45 (3): 499–525.

Goldman, Marshall I. 1980. *The Enigma of Soviet Petroleum: Half Empty or Half Full?* London: George Allen and Unwin.

Guoxing, Ji. 1998. "China versus South China Sea security." *Security Dialogue* 29 (1): 100–112.

Harrison, Selig S. 2002. "Quiet struggle in the East China Sea." *Current History* 101 (656): 271–277.

Heinberg, Richard. 2003. *The Party's Over: Oil, War, and the Fate of Industrial Societies*. Gabriola Island, BC: New Society Publishers.

Homer-Dixon, Thomas F. 1999. *Environment, Scarcity, and Violence*. Princeton, NJ: Princeton University Press.

Ikenberry, John G. 1988. *Reasons of State: Oil, Politics, and the Capacity of the American Government*. Ithaca, NY: Cornell University Press.

Jaffe, Amy Myers, and Robert A. Manning. 1998. "The myth of the Caspian 'Great Game': The real geopolitics of energy." *Survival* 40 (4): 112–129.

Jervis, Robert. 2002. "Theories of war in an era of leading power peace." *American Political Science Review* 96 (1): 1–14.

Jinming, Li and Li Dexia. 2003. "The dotted line on the Chinese map of the South China Sea." *Ocean Development and International Law* 34 (1–2): 287–295.

Johnson, Robert H. 1989. "The Persian Gulf in U.S. strategy: A skeptical view." *International Security* 14 (1): 122–160.

Johnston, Alastair I. 1995. *Cultural Realism: Strategic Culture and Grand Strategy in Chinese History*. Princeton, NJ: Princeton University Press.

Johnston, Alastair I. 2003. "Is China a status quo power?" *International Security* 27 (4): 5–56.

Kang, Daniel. 2003. "Getting Asia wrong: The need for new analytical frameworks." *International Security* 27 (4): 57–85.

Kaysen, Carl. 1990. "Is war obsolete? A review essay." *International Security* 14 (4): 42–64.

Kegley, Charles W., Jr. 1993. "The neoidealist movement in international studies? Realist myths and the new realities." *International Studies Quarterly* 37 (2): 131–146.

King, Gary, Robert O. Keohane, and Sidney Verba. 1994. *Designing Social Inquiry: Scientific Inference in Qualitative Research*. Princeton, NJ: Princeton University Press.

Kissinger, Henry. 1982. *Years of Upheaval*. Boston: Little, Brown.

Klare, Michael T. 2001. *Resource Wars: The New Landscape of Global Conflict*. New York: Metropolitan Books.

Klare, Michael T. 2004. "Geopolitics reborn: The global struggle over oil and gas pipelines." *Current History* 103 (677): 428–433.

Klare, Michael T. 2006. "Fueling the dragon: China's strategic energy dilemma." *Current History* 105 (690): 180–185.

Kleveman, Lutz. 2003. *The New Great Game: Blood and Oil in Central Asia*. New York: Atlantic Monthly Press.

Kleveman, Lutz. 2004. "Oil and the new Great Game." *Nation* 278 (6): 11–14.

Kober, Stanley. 2000. "The Great Game, Round 2: Washington's misguided support for the Baku-Ceyhan oil pipeline." *CATO Foreign Policy Briefing* 63. Washington, DC: CATO Institute.

Krasner, Stephen D. 1978. *Defending the National Interest: Raw Materials Investments and U.S. Foreign Policy*. Princeton, NJ: Princeton University Press.

Kurlantzick, Joshua. 2007. *Charm Offensive: How China's Soft Power Is Transforming the World*. New Haven, CT: Yale University Press.

Layne, Christopher. 2004. "China's role in American grand strategy: Partner, regional power, or great power rival?" In *The Asia-Pacific: A Region of Transitions*, ed. Jim Rolfe, 54–80. Honolulu: Asia-Pacific Center for Security Studies.

Leifer, Michael. 1995. "Chinese economic reform and security policy: The South China Sea connection." *Survival* 37 (2): 44–59.

Lewis, John W., Hua Di, and Xue Litai. 1991. "Beijing's defense establishment: Solving the arms-export enigma." *International Security* 15 (4): 87–109.

Liberman, Peter. 1998. *Does Conquest Pay? The Exploitation of Industrialized Societies*. Princeton, NJ: Princeton University Press.

Lieven, Anatol. 1999/2000. "The (not so) Great Game." *National Interest* 22 (58): 69–80.

Mandelbaum, Michael. 1998. "Is major war obsolete?" *Survival* 40 (4): 20–38.

Mandelbaum, Michael. 2002. *Ideas That Conquered the World: Peace, Democracy, and Free Markets in the Twenty-first Century*. New York: Public Affairs.

Manning, Robert. 2000a. *The Asian Energy Factor: Myths and Dilemmas of Energy, Security, and the Pacific Future*. New York: Palgrave.

Manning, Robert. 2000b. “The Asian energy predicament.” *Survival* 42 (3): 73–88.

Matthews, Ken. 1993. *The Gulf Conflict and International Relations*. New York: Routledge.

McNaugher, Thomas. 1985. *Arms and Oil: U.S. Military Strategy in the Persian Gulf*. Washington, DC: Brookings Institution Press.

Mearsheimer, John J. 1994/1995. “The false promise of international institutions.” *International Security* 19 (3): 5–49.

Mearsheimer, John J. 2002. “Liberal talk, realist thinking.” *University of Chicago Magazine* 94 (3). Available at <http://magazine.uchicago.edu/0202/features/index.htm> (accessed March 12, 2008).

Mearsheimer, John J., and Zbigniew Brzezinksi. 2005. “Clash of the titans.” *Foreign Policy* 146: 46–50.

Medeiros, Evan S., and Taylor M. Fravel. 2003. “China’s new diplomacy.” *Foreign Affairs* 82 (6): 22–35.

Misra, Amalendu. 2001. “Shanghai 5 and the emerging alliance in Central Asia: The closed society and its enemies.” *Central Asian Survey* 20 (3): 305–321.

Mojthed-Zadeh, Pirouz. 2000. “The geopolitics of the Caspian region.” In *The Caspian Region at a Crossroad: Challenges of a New Frontier of Energy and Development*, ed. Hooshang Amirahmadi, 175–186. New York: St. Martin’s Press.

Mueller, John. 1989. *Retreat from Doomsday: The Obsolescence of Major War*. New York: Basic Books.

Mueller, John. 1995. *Quiet Cataclysm: Reflections on the Recent Transformation of World Politics*. New York: HarperCollins.

Mueller, John. 2004. *The Remnants of War*. Ithaca, NY: Cornell University Press.

Mulvenon, James. 2007. “Rogue warriors? A puzzled look at the Chinese ASAT Test.” *China Leadership Monitor* 20. Stanford, CA: Hoover Institution.

Nye, Joseph S. 1980. “Energy nightmares.” *Foreign Policy* 40: 132–154.

Odell, Peter. 2004. *Why Carbon Fuels Will Dominate the 21st Century’s Global Energy Economy*. New York: Multi-Science Publishing.

Odom, William E., and Robert Dujarric. 1995. *Commonwealth or Empire? Russia, Central Asia, and the Transcaucasus*. Indianapolis: Hudson Institute.

O’Hanlon, Michael. 2001. “Coming conflicts: Interstate war in the new millennium.” *Harvard International Review* 23 (2): 42–46.

Olcott, Martha Brill. 2001. “Revisiting the twelve myths of Central Asia.” Working Paper 23. Washington, DC: Carnegie Endowment for International Peace.

Orme, John. 1997/1998.“The utility of force in a world of scarcity.” *International Security* 22 (3): 138–167.

Oxman, Bernard H. 1996. “Caspian Sea or lake: What difference does it make?” *Caspian Crossroads* 1 (4): 1–13.

Pegg, Scott. 2003. "Globalization and natural resource conflicts." *Naval War College Review* 56 (4): 82–96.

Pollack, Kenneth M. 2004. *The Persian Puzzle: The Conflict between Iran and America*. New York: Random House.

Rabinowitz, Philip D., Mehdi Z. Yusifov, Jessica Arnoldi, and Eyal Hakim. 2004. "Geology, oil and gas potential, pipelines, and the geopolitics of the Caspian Sea region." *Ocean Development and International Law* 35 (1): 19–40.

Ray, James L. 1989. "The abolition of slavery and the end of international war." *International Organization* 43 (3): 405–439.

Reich, Bernard. 1987. *The Powers in the Middle East: The Ultimate Strategic Arena*. New York: Praeger.

Roberts, John. 1996. *Caspian Pipelines*. London: Royal Institute of International Affairs.

Roberts, Paul. 2004. *The End of Oil: On the Edge of a Perilous New World*. New York: Houghton Mifflin.

Rosecrance, Richard. 1986. *The Rise of the Trading State: Commerce and Conquest in the Modern World*. New York: Basic Books.

Rosecrance, Richard. 1999. *The Rise of the Virtual State: Wealth and Power in the Coming Century*. New York: Basic Books.

Ross, Dennis. 1981. "Considering Soviet threats to the Persian Gulf." *International Security* 6 (2): 159–180.

Ross, Robert S. 1999. "The geography of the peace: East Asia in the Twenty-first century." *International Security* 23 (4): 81–118.

Ruseckas, Laurent. 1998. "State of the field report: Energy and politics in central Asia and the Caucasus." *Access Asia Review* 1 (2): 41–84.

Salameh, Mamdouh. 1995. "China, oil, and the risk of regional conflict." *Survival* 37 (4): 133–146.

Salameh, Mamdouh G. 2001. "A third oil crisis?" *Survival* 43 (3): 129–144.

Schlesinger, James. 1990. "Oil and power in the nineties." *National Interest* 19: 113–115.

Shaffer, Brenda. 2005. "From pipedream to pipeline: A Caspian success story." *Current History* 104 (684): 343–346.

Shambaugh, David. 1999/2000. "China's military views the world: Ambivalent security." *International Security* 24 (3): 52–79.

Shambaugh, David. 2004/2005. "China engages Asia: Reshaping the regional order." *International Security* 29 (3): 64–99.

Shambaugh, David. 2005. "China's military modernization: Making steady and surprising progress." In *Strategic Asia 2005–06: Military Modernization in an Era of Uncertainty*, ed. Ashley J. Tellis and Michael Wills, 67–104. Washington, DC: National Bureau of Asian Research.

Simmons, Mathew R. 2006. *Twilight in the Desert: The Coming Saudi Oil Shock and the World Economy*. New York: Wiley.

Song, Yann-Huei. 2003. "The overall situation in the South China Sea in the new millennium: Before and after the September 11 terrorist attacks." *Ocean Development and International Law* 34 (1–2): 229–277.

Starr, Frederick S. 1997. "Power failure: American policy in the Caspian." *National Interest* 47: 20–31.

Storey, Ian J. 1999. "Creeping assertiveness: China, the Philippines, and the South China Sea dispute." *Contemporary Southeast Asia* 21 (1): 95–115.

Talbott, Strobe. 1997. A farewell flashman: American policy in the Caucasus and Central Asia. Speech delivered at the Central Asia Institute of the Paul H. Nitze School of Advanced International Studies of Johns Hopkins University, July 21.

Telhami, Shibley. 1992. "Between theory and fact: Explaining American behavior in the Gulf." *Security Studies* 2 (1): 96–121.

Thao, Nguyen Hong. 2003. "The 2002 Declaration on the Conduct of Parties in the South China Sea: A note." *Ocean Development and International Law* 34 (1–2): 279–285.

Thompson, Scott W. 1982. "The Persian Gulf and the correlation of forces." *International Security* 7 (1): 157–180.

Tucker, Robert W. 1975a. "Further reflections on oil and force." *Commentary* 59 (3): 45–56.

Tucker, Robert W. 1975b. "Oil: The issue of American intervention." *Commentary* 59 (1): 21–31.

Tucker, Robert W. 1980–1981. "The purposes of American power." *Foreign Affairs* 59 (2): 241–274.

Valencia, Mark J., and Yoshihisa Amae. 2003. "Regime building in the East China Sea." *Ocean Development and International Law* 34 (2): 189–208.

Valenta, Jiri. 1980. "From Prague to Kabul: The Soviet style of invasion." *International Security* 5 (2): 114–141.

Waldron, Arthur. 1997. "How not to deal with China." *Commentary* 103 (3): 44–49.

Waltz, Kenneth N. 1979. *Theory of International Politics*. Reading, MA: Addison-Wesley.

Waltz, Kenneth N. 1981. "A strategy for the rapid deployment force." *International Security* 5 (1): 49–73.

Whiting, Allen S. 2001. "China's use of force, 1950–96, and Taiwan." *International Security* 26 (2): 103–131.

Wohlforth, William C. 2002. "U.S. strategy in a unipolar world." In *America Unrivaled: The Future of the Balance of Power*, ed. John G. Ikenberry, 98–118. Ithaca, NY: Cornell University Press.

Wohlforth, William C. 2004. "Revisiting balance of power theory in central Eurasia." In *Balance of Power: Theory and Practice in the 21st Century*, ed. T. V. Paul, James J. Wirtz, and Michel Fortmann, 214–238. Palo Alto, CA: Stanford University Press.

Woodrow Wilson International Center for Scholars. "Soviet invasion of Afghanistan." Cold War International History Project. Virtual archive 2. Available at <http://www.wilsoncenter.org/index.cfm?topic_id=1409&fuseaction=va2.browse&sort=Collection&item=Soviet%20Invasion%20of%20Afghanistan> (accessed November 16, 2009).

Wu, Samuel S. G., and Bruce Bueno de Mesquita. 1994. "Assessing the dispute in the South China Sea: A model of China's security decision making." *International Studies Quarterly* 38 (3): 379–403.

Yergin, Daniel. 1991. *The Prize: The Epic Quest for Oil, Money, and Power*. New York: Simon and Schuster.

Yergin, Daniel, Dennis Elkof, and Jefferson Edwards. 1998. "Fueling Asia's recovery." *Foreign Affairs* 77 (2): 34–50.

Zha, Daojiong, and Mark J. Valencia. 2001. "Mischief Reef: Geopolitics and implications." *Journal of Contemporary Asia* 31 (1): 86–103.

10

Responses to Alternative Forms of Mineral Scarcity: Conflict and Cooperation

Deborah J. Shields and Slavko V. Šolar

Humans have used mineral resources since prehistoric times. People with access to flint were more successful hunters. People who could build sturdy shelters were protected from the elements and potentially from enemies as well. Those with stores of precious metals had an advantage in trade. Over time, the technologies of mineral exploration, extraction, processing, and use have advanced. Both the variety of resources exploited and the applications to which they are put have expanded dramatically.

Today, minerals are essential inputs to economic systems, the driving force for some local, regional, and national economies, and the basis of the built environment. Governments have an interest in ensuring a stable supply of minerals to support their economies and provide for the defense of their nations, and throughout history have been willing to use a variety of means to do so. This has led to conflict in some cases and cooperation in others.

A major driver of mineral-related conflicts and cooperative actions is real or perceived scarcity. Malthusian fears about the adequacy of resource stocks have been revived in recent years, partly in reaction to the rising world population, which is predicted to reach eight billion people by 2025. Of most pressing concern with respect to minerals are the expanding economies of high-population, developing nations such as India and China. The consumption of resources increases as incomes rise and the middle class grows larger in these countries. Figure 10.1 below illustrates recent trends in mineral resource consumption in China.

Dianzuo Wang (2005) predicts that the growth in demand for minerals in China will increase rapidly for the next fifteen to twenty years, with the annual consumption of steel reaching 330 million tons as early as 2015. John DeYoung reports that China's consumption of copper in 2020 could equal the level of worldwide consumption in the mid-1990s

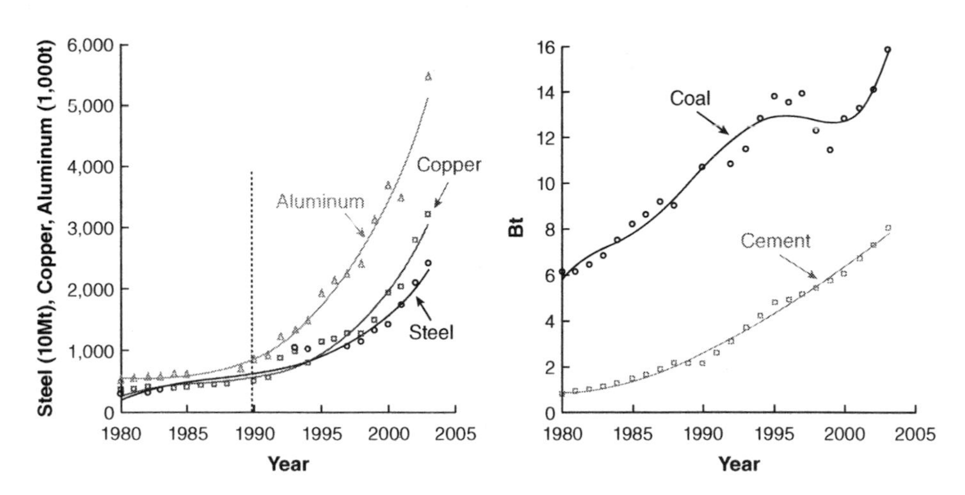

Figure 10.1
Trends in resource consumption in China, 1980–2005 (Wang 2005). *Note*: 10 Mt = 10 thousand tons; 1000 t = thousand tons; Bt = billion tons

(quoted in Darmstadter 2001). Increased demand has dramatically affected mineral prices. In 2006, the price of copper (in US$) was $2.25 per pound, the price of lead was $0.50 per pound, and the price of gold was $550 an ounce (Andrews 2008). By April 2008, those prices had reached $4.35 per pound, $1.45 per pound, and $947 per ounce, respectively (LME 2008; GFMS 2008). By December 2008, however, prices had dropped to $1.41 per pound, $0.44 per pound, and $810 per ounce, respectively (LME 2008; GFMS 2008), reflecting the decreased demand stemming from the turmoil in financial markets and concerns about the global recession. Economic downturns do not last indefinitely, though; demand has already begun to rebound.

A portion of China's mineral commodity consumption is used in the production of goods that are subsequently exported. Thus, decreased demand for manufactured goods in developed countries could temper the rate of increase in China's mineral consumption, at least in the short term. A reduction in export earnings could also negatively impact China's booming construction sector. In response, the Chinese government announced plans for an economic stimulus package of US$586 billion, which will largely be directed toward infrastructure projects. Regardless of current market fluctuations, in the long term Chinese consumption is expected to rise even further than it already has. Experts predict similar increases in India (Menzie 2006). Moreover, the increased demand for minerals in the developing world is not being offset by comparable decreases in consumption levels in developed nations (Moll, Bringezu, and Schütz 2005).

In a keynote speech to the World Mines Ministries Forum, Somit Varma (2008) stated that "high prices and perceptions of scarcity have made it clearer that meeting growing demand is not easy, but requires secure access to resources and substantial, timely investment." It will also require the discovery of new deposits to replace those being depleted. Yet serious technical challenges will need to be overcome if this is to happen (Skinner 2001). Under these circumstances, it is not surprising to see the specter of scarcity rising once again. Governments and industry in Europe, North America, Asia, and elsewhere are expressing concern about the availability of resources as well as the security of mineral supply. Many are investing in commodity-producing firms or acquiring mining concessions in other (often poor, developing) nations, in what some authors see as a repeat of the Great Game (Rajan 2006).[1]

Predictions or perceptions of scarcity can increase the potential for interstate conflict, either between producing and consuming countries, or among consuming countries, as each player tries to secure their position in a dynamic mineral market. Conversely, mineral scarcity can motivate interstate cooperation when nations see it in their self-interest to do so. Perceived or real physical scarcity is not the only driver of decisions regarding minerals. Although having access to enough mineral material was of singular importance at one time, what societies consider to be within the scope of concern vis-à-vis mineral production has broadened. The manner in which minerals are produced, where they originate, and who is impacted by their production also all matter at present to some stakeholders and governments (Shields and Šolar 2006). Figure 10.2 illustrates this progressive change.

Initially, discovery and access were the drivers of mineral-related decisions made by those in power (tribal leaders, kings, or governments). During the Industrial Revolution, property rights, investment, and access to capital rose in importance. By the late industrial age, the role of workers and their rights began to be recognized. With the birth of the environmental movement in the 1960s and 1970s, the impacts of mining and mineral processing on water, air, and land started to be addressed. Capital investment, workplace safety, mine reclamation, and labor rights are now generally accepted as relevant issues among most, though not all, governments and within large segments of the mining industry.

Furthermore, the emergence and embrace of the sustainable development and corporate social responsibility paradigms in the late twentieth

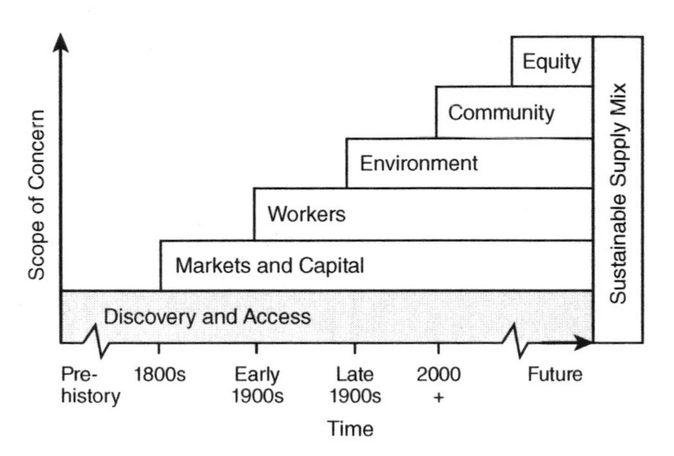

Figure 10.2
Expansion of the scope of concern about the mineral supply (Shields and Šolar 2006)

and early twenty-first centuries mark two other significant expansions of the scope of concern about the mineral supply. There is an increasing acknowledgment that communities need to be consulted about business activities that have the potential to impact their social structure, health, and economy, albeit with only partial consensus on how this should be accomplished. And there is a growing recognition that equity within and across generations needs to be considered in the production, use, and disposal of minerals as well as in the reclamation and aftercare of mining sites. Any of the issues outlined above can become a source of interstate conflict if the two parties involved in a mineral transaction do not agree on how they should be handled. Not all mineral-producing nations place worker health and safety or environmental protection at the forefront of concern, nor do they actively address the intergenerational equity aspects of mineral consumption or production.

When two states have fundamentally different points of view on such issues, conflicts will arise if one party feels that their position must prevail or be imposed on the other side. Conflict also can stem from unfulfilled expectations among partners in a mineral transaction. In a world of global mineral trade and multinational mining companies (some with a net worth exceeding that of the nation where they are mining), a high percentage of mineral activities and related conflicts can be treated as internationalized because the parties involved have different countries of origin. For example, conflicts can arise between a multinational mining firm and either the community near the mine or the host government, related to environmental practices, community impacts, or the distribution of economic rents. In some cases participants in such disputes look to their respective governments for redress, potentially turning an internationalized conflict into an interstate conflict. If, on the other hand, the parties to a mineral transaction act in good faith, practice open and forthright communication, and engage in dialogue and negotiation, disagreements can be dealt with in ways that prevent interstate conflict. Moreover, cooperation may take place in these contexts when the participants recognize the value of applying the principles of sustainable development and corporate social responsibility to mineral extraction and trade, as will be discussed later in this chapter.

Other examples of interstate cooperation have been observed in recent years. Producing and consuming nations have recognized the need to coordinate their efforts in the areas of information gathering and the assessment of minerals resources, which will contribute to a stable supply. Increasing scarcity has also motivated consuming nations to begin discussing the need to synchronize their national mineral policies. Producing nations have likewise come together to facilitate trade and investment in their minerals industry and strengthen their regional capacities. Finally, nations have increased their cooperation through intergovernmental forums, such as the arbitration of seabed mineral rights under the United Nations Convention on the Law of the Sea (UNCLOS) passed in 1982.[2]

The purpose of this chapter is to consider mineral scarcity and how it relates to interstate conflict and cooperation. We will address three core questions: What is meant by the term mineral scarcity, and by extension, are minerals scarce? Is conflict over minerals caused by real or perceived resource scarcity, or by other factors? How does cooperation, based on scarcity and other related issues, manifest itself in the minerals sector among the respective parties? We begin with a discussion of what scarcity means and how mineral scarcity has been characterized in the literature. Then we distinguish between physical, situational, political, and social scarcity. Following that, we turn to mineral conflicts, describing different participants and motivations as well as identifying whether scarcity is a core driver. The next section of the chapter addresses the concepts of intraindustry, international, and intergovernmental cooperation and mutual benefit. We talk about historic and current drivers of cooperation—primary among these being sustainable development—and then describe a variety of ongoing activities. We end with thoughts on real and perceived scarcity, and consider whether cooperation in its various forms is a likely and feasible long-term policy choice.

Minerals Scarcity

Minerals are found in varying forms, concentrations, and amounts throughout the earth's crust. Once a mineral is considered useful to humans and technology has advanced to the point that the mineral can be processed, it is relabeled as a resource. When minerals become resources, their abundance and availability—that is, their supply—become important. Abundance is a question of geologic and economic stocks. In contrast, availability is a question of flows through the mineral system.

Minerals become scarce for two basic reasons: the depletion of stocks or the disruption of flows. The latter form of scarcity could be the result of depletion, although to date there have been no instances of long-term scarcity due to depletion. Rather, for the purposes of this chapter, flow disruption refers to such situations as a temporary cessation of production, demand that exceeds production capacity, or political actions such as embargoes. Minerals are traded in domestic and international markets, and so regardless of the underlying reason, when mineral flow through the supply system is inadequate to meet demand, prices rise. Markets respond in multiple ways: exploration for more resources, research into new production technologies, substitution of alternative materials, recycling, and more efficient use (Shields, Šolar, and Miller 2007). Figure 10.3 depicts this process in a market economy.

One short-term governmental response to scarcity would be releases from existing national stockpiles. Responses to price increases could include price controls, subsidies to consumers, or tax reductions for industry. Intermediate-term responses could include support for research, innovation, exploration, and the diversification of sources. In extreme cases, governments resort to force to ensure a steady mineral flow. In recent years, however, there has been an increasing reliance on interstate cooperative solutions to problems of scarcity.

The short-term industry response will include the production of lower-grade ores—as price rises, the cutoff grade decreases. Intermediate-term responses include production from marginally economic deposits. Firms will increase exploration, and new supplies will enter the market as that exploration leads to the discovery of new deposits and as innovation increases the viability of known and newly found deposits. In the short term, the demand may be inelastic. Yet the intermediate response will be conservation, with an increase in substitution, reuse, and recycling.

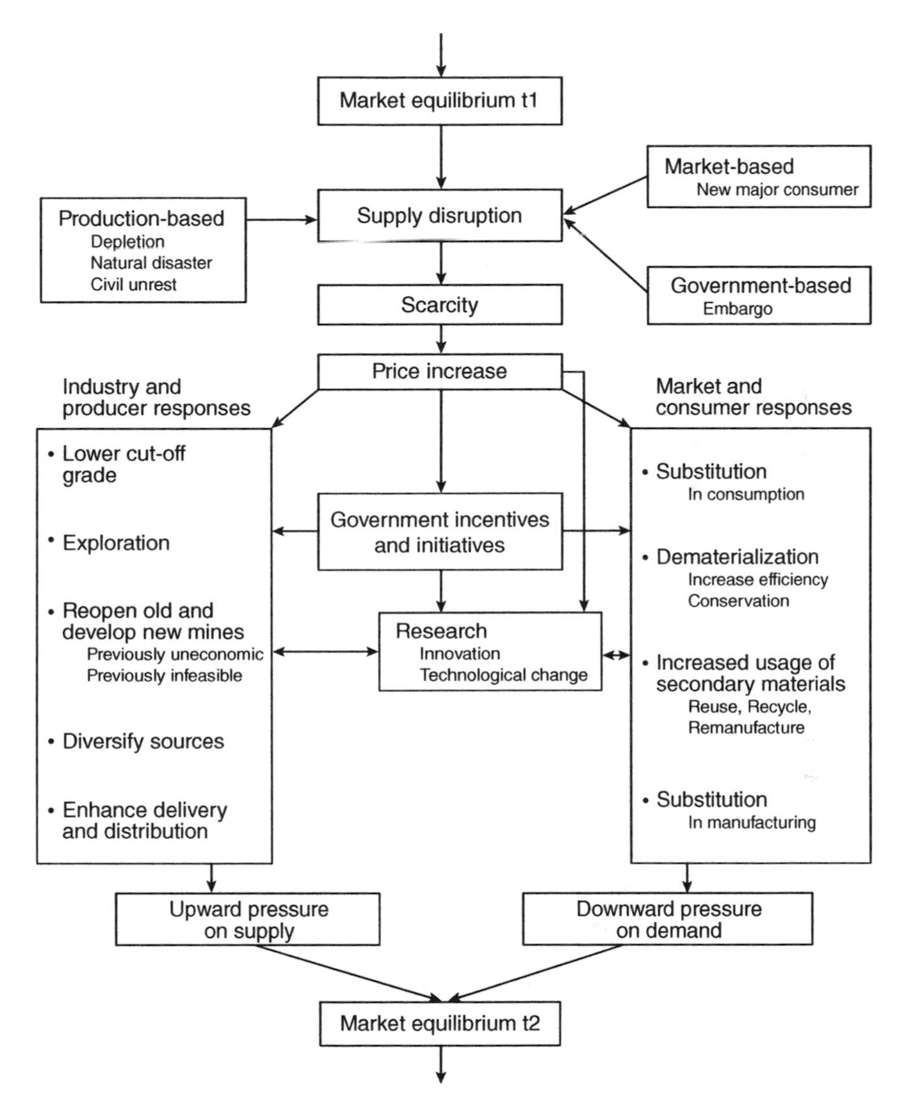

Figure 10.3
Market response to scarcity model for minerals (extended after Shields, Šolar, and Miller 2007)

Research and innovation will support and magnify these trends. Over time, an equilibrium will be reestablished, though conceivably at a different price, and the cycle will begin again.[3]

As noted above, scarcity can have multiple causal factors and can take several forms. Below, we investigate physical scarcity, which is a stock concept, in geologic and economic contexts. We then turn to situational, political, and social scarcity, which concern issues of flow through the minerals system. In these latter cases, geologic constraints are only relevant because mineral commodities are not distributed evenly across the globe; circumstances other than geology lead to the actual or perceived scarcity.

Physical Scarcity

The earth is made up of more than ninety chemical elements, some of which are geochemically abundant and others of which are geochemically scarce. The first group comprises twelve elements that account for 99.23 percent of the mass of the earth's continental crust. Ten are widely used: aluminum, iron, magnesium, manganese, silicon, calcium, sodium, potassium, titanium, and phosphorus. The remaining 0.77 percent of the crust is composed of the remaining elements, including metals of economic importance such as copper (Ayers 2001).

The exact amount and location of geochemically scarce elements cannot be known with certainty because they are hidden beneath the earth's surface. Geologists have made great strides identifying the subsurface characteristics that make the discovery of specific minerals more likely. Still, the geologic processes that lead to deposition differ across minerals, making it inappropriate to generalize about location or abundance (Skinner 2001). Concentrations of minerals of economic interest are called deposits. Deposits are said to contain ores; everything else is termed common rock. In some instances, minerals exist in common rock at ever-decreasing concentration, down to the level of atomic substitution.

The McKelvey (1972) diagram (figure 10.4) is used as a way to classify ores into two categories—resources and reserves—and the amounts reported are estimates of the long- and short-term abundance of various minerals. Resources are defined as concentrations of material from which it is feasible to economically extract a mineral commodity. Resources may be known—that is, identified—or as yet undiscovered. Because of the uncertainty inherent in assessing the unknown, resource amounts are

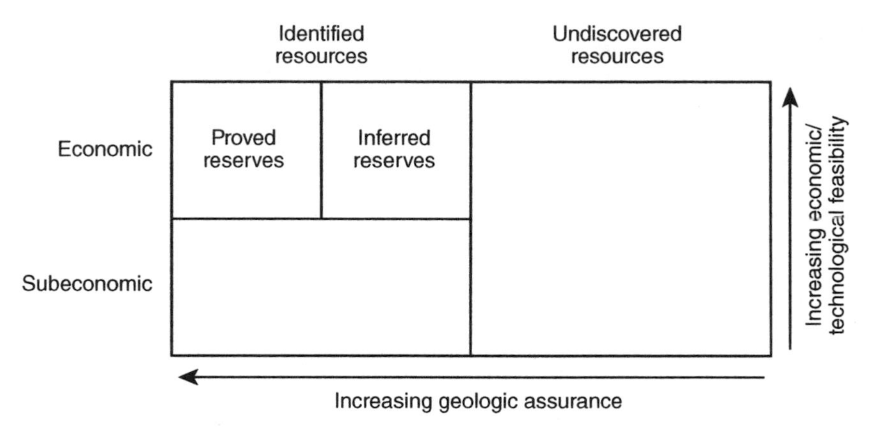

Figure 10.4
Mineral resource classification (after McKelvey 1972)

often presented as probability distributions. Reserves are a subset of identified resources that are currently profitable to exploit given existing technology and prices.

Reserves do not necessarily exist forever simply because they are known. If commodity price falls below break-even costs, reserves cease to exist and return to the resources category. The recent rise and subsequent fall in mineral prices has had this affect on reported reserves; such changes tell us nothing about the actual abundance (or scarcity) of a mineral, though. For example, the price of nickel dropped 80 percent between the beginning and end of 2008 (GFMS 2008), but the known resource and the stock of nickel in the earth's crust changed relatively little. Moreover, technological advances in extraction and processing can transform common rock into a mineral resource. Such progress and/or higher prices allow subeconomic resources to move (upward) into economic resources or recoverable reserves (Darmstadter 2001). Technologies do not currently exist to economically produce minerals from common rock when they are distributed at the level of atomic substitution.

There are two perspectives on the physical scarcity of minerals: the fixed stock and opportunity cost paradigms (Tilton 2003). The fixed stock paradigm takes a geologic perspective. It starts from the premise that the supply of any mineral commodity is finite. Continued extraction and use will lead first to scarcity and eventually to exhaustion (Baumol and Blackman 2007). A mineral would be considered scarce if the majority of ore deposits were believed to have been found, the probability of new discoveries was deemed low, or new resources would have to come from common rock. It could also mean that undiscovered resources (the right-hand side of the McKelvey diagram) are assumed to "lie beneath a cover of younger barren rock and are sufficiently deep so that today's prospecting technology cannot detect them" (Skinner 2001, 8). The fixed stock paradigm can serve as a starting point for analysis in cases where the uses of a mineral resource are largely or entirely dissipative and not substitutable—for example, phosphorous in agriculture (Goeller 1972). Its sole application is not appropriate where substitution, recycling, and technological innovation have the potential to offset depletion (NREL 2004).

Physical measures of scarcity are associated with the fixed stock paradigm. One commonly used measure is the static reserve index—that is, the ratio of reserves to current annual consumption. The result is an estimate of the number of years that reported reserves would last. John Tilton (2003) modified the measure to have production from reserves increase annually at different rates (0, 2, and 5 percent), which at the upper growth rate resulted in many mineral commodities becoming depleted in tens to hundreds of years. Two important assumptions underlie the static reserve index: constant consumption rates and no new discoveries, which together limit the meaningfulness of the measure.

A ratio of reserves to forecasted consumption or demand could be used instead. Such forecasts are challenging undertakings because of the need to simultaneously forecast population, technological change, rates of substitution, and rates of secondary production and recycling. Even assuming that the forecasts were of the highest quality, this measure of scarcity still would suffer from a fatal flaw; reserves are merely firms' working stocks rather than measures of the full fixed stock of the mineral. Firms seldom, if ever, completely delineate a deposit's reserves, particularly if decades of reserves are already known to exist. Moreover, the tax consequences for reporting reserves could provide an economic incentive limiting exploration and reporting. And as noted earlier, the amount of reserves can be directly affected by price changes.

The fixed stock view informed a number of books in the 1960s and 1970s, most famously the Club of Rome report, *Limits to Growth* (Meadows et al. 1972). The authors used an exponential growth model for the economy coupled with arithmetically increasing reserves to calculate the remaining years of life expectancy for a variety of minerals. Their methodology and results were strongly criticized.

Limits to Growth: The 30 Year Update was published in 2004 (Meadows, Randers, and Meadows 2004). The basic premise that consumption of the earth's resources at a high rate cannot be sustained without causing social, environmental, and economic collapse remained, as did concerns about exponential growth and imbalance between sources and sinks, all of which are valid points. The authors took a more hopeful tone than previously, however, suggesting that humanity could step back from the brink of disaster if appropriate actions are taken. That said, representatives of the Club of Rome have continued to predict impending exhaustion for some minerals (Schauer 2007), a position disputed by the U.S. Geological Survey (Menzie, Singer, and DeYoung 2005).

Shifting from reserves to resources in these types of measures does lead to greater life expectancies, but it does not remedy the core problem that probabilistic estimates do not reflect true availability. Conversely, use of the resource base, which is an estimate of ultimate crustal abundance, increases life expectancies to such a point that scarcity within our or our children's lifetimes ceases to be a concern for many commodities (Tilton 2003).

An alternative approach to physical scarcity is to observe the rate at which known and depleting deposits are being replaced through exploration and new discoveries. Robert Gordon, Bertram McInnes, and Thomas Graedel (2006) examined the per capita use of copper relative to an estimate of the size of the copper resources accessible in the earth's crust. They showed that discoveries of new deposits of copper ore have not kept pace with extraction or losses due to dissipation and disposal. The authors concluded that if the historic trends of discovery and extraction continue, scarcity will be the eventual result.

In contrast to the fixed stock approach, the opportunity cost paradigm has an economic perspective. As figure 10.3 shows, as a mineral commodity becomes less available, its price will rise, leading to more exploration, which increases the likelihood of discovery of new deposits. An increasing price also makes previously uneconomic ores or deposits commercially viable. Proponents of the opportunity cost paradigm further argue that the true physical exhaustion of a mineral is not possible because before that could happen, the price would rise to such a level that recycling would increase and eventually substitutes would be found. Ultimately, the availability of a mineral will be determined by what people are willing to give up for it—that is, its opportunity cost. This paradigm is widely applicable, and is particularly relevant for resources that are being recycled, have known substitutes, or are the subject of intensive research.

Economic measures of scarcity are more consistent with the opportunity cost paradigm than the fixed stock paradigm. These include price, marginal extraction cost, and user cost. The basic first-order condition for optimal resource extraction is:

$$P = C_q + \lambda$$

Where

P = extracted resource price

C_q = marginal extraction cost

λ = user cost

Price reflects what buyers are willing to pay for the mineral resource in question, and incorporates information about demand and expectations concerning the future. Many mineral prices declined in real terms from the mid-1800s to the end of World War II, but then rose until the early 1980s. Margaret Slade (1982) hypothesized that this U-shaped price curve was demonstrative of a situation in which increasing scarcity exerted a greater influence on the price than did technological progress. Increasing scarcity drives prices up because it is assumed that lower-cost deposits have already been identified and exploited. Prices then declined again until the relatively short-lived spike (three to five years depending on the commodity) that ended rather dramatically in fall 2008.

As Tilton (2003) has pointed out, price is an imperfect measure of scarcity. Price increases may reflect a variety of events that affect the flow of a commodity through the mineral system, rather than actual increasing scarcity—for example, events such as strikes, embargoes, or natural disasters like hurricanes. The price can also be affected by market imperfections, the actions of speculators and hedge funds, and government actions that alter market functioning. Moreover, it is widely recognized that most mineral prices do not reflect the full cost of production, since many environmental and social costs are not included in financial calculations (MMSD 2002). All of the foregoing suggests that the price is not a completely reliable indicator of increasing scarcity of a mineral. Further, in a recent paper, Peter Svedberg and Tilton demonstrated that the deflator used to calculate the real mineral price has a statistically significant affect on the level of the real price and its trend over time. They concluded that "real resource prices provide less support than widely assumed for the hypothesis that resources are becoming [either] more available or less scarce over time" (Svedberg and Tilton 2006, 501).

Marginal extraction costs have their own set of problems as predictors of increasing scarcity. C_q is the sum of capital and labor (and increasingly, at least part of the social and environmental) costs of producing a unit of mineral output. The marginal cost decreased for many years as technological progress generated more cost-effective exploration, extraction, and processing methods (Krautkraemer 2005). Factor substitution and the discovery of new high-grade deposits also drove costs downward in some cases. Recent cost increases have at least in part reflected the competition in markets for skilled labor, capital equipment, and materials. In such circumstances, it is difficult to know with high precision what percent of a cost increase has been caused by increasing scarcity.

We turn finally to user cost, also referred to as scarcity rent or Hotelling rent. User cost is the difference between market price and marginal extraction cost, and reflects the value of the foregone opportunity to use a nonrenewable resource in the future as well as the value of remaining stocks in the ground. As such it is a measure of scarcity, assuming that costs are constant. The difficulty is that user cost is actually a measure of beliefs about scarcity rather than an actual measure of scarcity. As Douglas Reynolds (1999) demonstrated, empirical data on costs and price cannot be depended on to indicate the true size of the resource endowment.

Harold Barnett and Chandler Morse (1963), in *Scarcity and Growth*, developed a scarcity index based on real unit costs and relative prices. They concluded that resource scarcity "had not yet, probably would not soon, and conceivably might not ever, halt economic growth" (Simpson, Toman, and Ayers 2005). In more recent years, the use of economic indicators of scarcity, and in particular price and cost, have been strongly criticized (Smith 1980; Norgaard 1990; Alonso et al. 2007), providing several motivations for the publication of *Scarcity and Growth Revisited* (Simpson, Toman, and Ayers 2005). The report concluded that the physical scarcity of mineral resources would not limit growth in most instances but instead that other issues might limit mineral availability (Menzie, Singer, and DeYoung 2005). The authors pointed to rising consumption in developing nations, the need for exploration in ever more inhospitable locations, shrinking exploration and research budgets, the lack of trained personnel, restrictions on mineral development, and environmental costs.

Tilton and Gustavo Lagos (2007) responded to the above-referenced Gordon, McInnes, and Graedel (2006) paper on copper stocks by noting that the fixed stock paradigm, while intuitively appealing, is flawed and overpessimistic (or in some instances, overly optimistic). They suggest that an opportunity cost perspective would be more useful in assessing copper availability, given that the U.S. Geological Survey has estimated that global copper resources are vast. Gordon, McInnes, and Graedel (2007) promptly responded, defending their work and stating that absolute abundance plays a role in the long-run availability of copper. Yet they acknowledged that factors such as legal, technological, environmental, and political constraints will also affect availability, and thus must be addressed. A recent National Research Council report (NRC 2008) takes a more overtly opportunity cost view, arguing that over the longer term, availability of minerals and mineral products will largely be a function of investment, and the various factors that influence the level of investment, as well as its geographic allocation and success.

Flow Disruption Scarcity

As the preceding discussions illustrate, the two sides of the scarcity debate are increasingly acknowledging the merits of each other's positions. Scarcity may be the result of limited or diminishingly accessible material in the earth's crust, although this is considered a long-term concept for most minerals. Alternatively, scarcity may be the result of a short- or medium-term imbalance between supply and demand—that is, the amount of a mineral commodity available at a specific location, at a certain point in time, is less than the amount desired. As noted in our market-flow diagram and explored above, such imbalances may be caused by a variety of constraints, virtually all of which can be overcome in time, assuming that society is willing to pay the costs to do so. We label these types of scarcity as situational, political, and social scarcity.

Situational Scarcity

Situational scarcity is caused by one or more of a broad set of circumstances that act to limit the flow of minerals to markets. The NRC report (2008, 22) identified the following as potential causes of a supply-demand imbalance: "(1) significant increase in demand; (2) thin markets; (3) concentration of production; (4) production predominantly as a byproduct; and (5) lack of available old scrap for recycling or of the infrastructure required for recycling." To these we add: a lack of trained professionals such as geologists, mining engineers, and miners; a level of investment in exploration too low to ensure discoveries of enough new deposits to replace those being depleted; inadequate throughput capacity resulting from a lack of investment in the maintenance of existing facilities and the construction of new facilities; a lack of investment, or discoveries, in basic and applied science, thereby restricting the opportunities for technological advancements; and constraints on mining imposed by either climate change, or the availability of energy or water. Locational scarcity, which is a subset of situational scarcity, is caused by either a lack of minerals in one specific geographic location or logistic bottlenecks, such as inadequate port capacity or a shortage of container ships. As we will demonstrate later, situational scarcity in the form of temporal and locational imbalances offer numerous opportunities for interstate cooperation, as does scarcity stemming from political or social choices.

Political Scarcity

Political scarcity occurs when the flow of a mineral commodity is halted or restricted due to choices made or actions taken by governments. The most obvious example of economically motivated political scarcity is an embargo. One producer or a cartel of producers acts to restrict the flow of a mineral from its country to one or more consuming countries with the goal of increasing the market price (as distinguished from restricting the flow so as to deny resources to an opponent). Countries electing to pursue this strategy must balance the potential for increased revenues from more expensive commodities against the potential for decreased demand as consumers respond to higher prices by choosing substitutes or increasing the rate of recycling. Another economically motivated, political restriction is the limiting of exports as a negotiating position over rent distribution. Nevertheless, creating a medium- or long-term scarcity in this manner would be self-defeating if the initiating nation were to lose more revenue over the period of restriction than they would subsequently gain from higher levels of retained economic rent.

The intentional restriction of mineral shipments to enemy states during wartime has historical precedents, such as when the Allied powers strove to limit shipments of energy and mineral commodities into Germany during World War II. During the Cold War, there was concern that the Soviet Union would attempt to deny the United States access to strategic and critical minerals either directly or through actions taken by proxy states. Consistent with this view, Gautam Sen (1984) maintained that controlling markets for manufactured goods and minerals has military significance. Robert Mandel (1998, 76) labeled this the strategic minerals threat: "a perceived threat rather than a concrete disruption or application of a resource weapon."

The potential for scarcity to be caused by strategic rivalry among global powers remains. Michael Klare (2002, 1) reframes this as an economic issue, contending that "whereas Cold War–era divisions were created and alliances formed along ideological lines, economic competition now drives international relations—and competition over access to [natural resources] has intensified accordingly." He predicts that there will be "a reconfigured cartography in which resource flows rather than political and ideological divisions constitute the major fault lines" (Klare 2002, 52).

If he is correct, then economic security may be used as a justification for several different types of actions that have the potential to create scarcity in world markets. The government of a mineral-producing nation could impose export restrictions, effectively funneling all domestic production to its own economy rather than to international markets. Alternatively, countries might assert exclusive control over the resources of a developing nation through licensing agreements or the outright acquisition of mines, with the intention of shipping all production to their home country, and thus reducing the volume of minerals available for world trade. China is taking both of these approaches. Finally, a nation could stake a territorial claim on the seabed to gain control of the mineral resources beneath it. To this end, Russia planted a flag on the seabed at the North Pole in 2007.

Politically inspired threats and risks were the impetus for the creation of the U.S. National Defense Stockpile (NDS) in 1939 (NMAB 2008). Over the NDS's life, materials have been removed from and added to it in response to changing technology, military needs, and strategic considerations. More than US$5.9 billion worth of materials were sold from the stockpile from fiscal year 1993 through fiscal year 2005. The importance of various minerals and the need for an NDS are now being reconsidered. The National Research Council convened the Committee on Assessing the Need for a Defense Stockpile to assess the need for and value of the NDS, current defense material needs, and the necessity of stockpiling. It concluded that there were three major threats to critical material supply, including an increased global demand for mineral commodities and materials, a diminished domestic supply and processing capability along with greater dependence on foreign sources, and a higher risk of and uncertainty about supply disruptions (NMAB 2008, 3).

A contemporaneous report (NRC 2008) utilized a criticality matrix approach to identify minerals that warrant concern. Criticality is defined as a function of the impact of supply restrictions and supply risk, which in turn is determined by the importance in use and availability. The report defined several categories. First, "minerals that are *essential* to the economy in certain applications and are yet *not critical*, at least at present, in that the risk of supply restriction is low" (NRC 2008, 20). Minerals likely to fall in this category of essential but not critical include copper, bauxite (the mineral raw material for aluminum), iron ore, and construction aggregates. Conversely, the minerals with the highest degree of criticality (of the eleven examined) were platinum group metals, rare earth oxides, indium, manganese, and niobium.[4]

Warring among factions within a country can also limit the flows of minerals out of that country. Whether this will result in global scarcity depends on the percentage of the world market share that the in-country mines produce. In the 1970s, Zaire (now the Democratic Republic of the Congo), together with neighboring Zambia, produced two-thirds of the world's cobalt. When internal strife disrupted production in Zaire, the resulting shortages drove the price of cobalt up 380 percent over a two-year period (Alonso et al. 2007) in reaction to fears that exports would cease for a period of time. A more generalized form of scarcity risk occurs when most production of a commodity takes place in politically unstable countries or in those whose governments are not friendly to import-dependent nations.

Theoretically, political scarcity could occur for an individual nation if it banned imports of a mineral from a specific producing country. This could be done for a variety of reasons including the desire to economically punish a nation for actions deemed inappropriate or unacceptable by the importing nation. For example, a country might ban the import of "conflict resources," the profits from which finance civil war or terrorist acts, or minerals from a country whose government commits atrocities against its own or other countries' citizens. We have found no instances in which a country pursued such a course to the point of creating actual, self-imposed scarcity and measurable economic harm.

Social Scarcity

Social scarcity may happen if citizens of a country decide that the environmental and social costs of extraction or processing are too great to bear. The expanded issues of concern regarding mineral supply (workers, the environment, communities and inter- and intragenerational equity) are drivers of this type of scarcity. Restrictions on exploration, the refusal to grant permits for mine development, or the imposition of regulations that make mines or processing facilities uneconomic could limit the mineral supply. Scarcity will ensue until or unless those supplies are replaced by minerals produced in another location, or until issues are resolved in a manner that the host society finds acceptable.

Finally, certain types of scarcity do not fit neatly into any of the above-named categories. Technological constraints imposed by the basic nature of materials, which are extremely difficult to overcome regardless of the amount of money expended on scientific research, is one such form of scarcity. Mineral scarcity may actually be driven by the scarcity of something else equally vital, yet still capable of limiting mineral activity. Specifically, the absorptive capacity of the environment, its ability to act as a sink for anthropogenic wastes, may ultimately place a finite boundary on the amount of mineral production or processing that can take place.

As will be shown in the following section, perceived impending scarcity, or actual scarcity due to situational, political, or social circumstances, are drivers of mineral conflict. Conversely, no conflicts in recent years have resulted from true physical scarcity—that is, the actual depletion of ultimate mineral stocks.

The Linkage between Minerals and Conflict

Conflict is defined as "the clashing of interests on national values of some duration and magnitude between at least two parties that find their interests incompatible, express hostile attitudes, or take action, which damages the other parties' ability to pursue those interests" (HIIK 2004). Minerals have frequently been associated with violent conflict, although in recent years these conflicts have not been driven by fears of depletion per se. A greater understanding of the distribution of mineral resources in the earth's crust and a globalized economy ensures that depletion of a single deposit does not imply a loss of access to the resource for any nation. Conflicts may be driven by the locational, political, or social aspects of scarcity, however, including those related to the extended scope of concern about mineral supply discussed in this chapter's introduction. The conflicts may be interstate, but are much more commonly internationalized because of the nature of the participants involved.

As Michael Ross (2003) points out, many recent conflicts did not start out being about minerals, but the presence of mineral wealth exacerbated existing problems and increased the likelihood of violence. Once that happens, the conflict becomes more difficult to resolve. Other authors have concurred that the control of minerals can become a focus of fighting that originally started for other reasons, that the wealth from their sale can finance continued fighting, and that when profits from mineral sales are used to buy arms, the intensity of the conflict may increase (AID 2004; Ballentine and Sherman 2003; Le Billon 2003; Ross 2002). Given the frequency of such disputes in recent years, we first consider the role of minerals in civil conflict and how these conflicts become internationalized. We then turn to other types of interstate and internationalized conflicts.

Abundance- and Scarcity-Driven Conflicts

A field of study that emerged in the late 1990s has focused on the nexus of mineral resources, violence, and civil war. The list of gem- and mineral-related civil wars is depressing in its litany of past and ongoing human suffering. Table 10.1 presents some recent conflicts. It is estimated, for example, that ongoing conflicts (many minerals related) in the Democratic Republic of the Congo have led to the deaths of more than four million people since 1997 from direct violence, casualties at unregulated artisanal mines, and conflict-induced starvation and disease (Global Witness 2006a).

Table 10.1
Recent civil wars (1990–2005) exacerbated by gems and minerals (Global Witness 2006a)

Country	Duration	Resources
Afghanistan	1978–2001	gems
Angola	1975–2002	diamonds
Burma	1949–	tin, gems
Cambodia	1978–1997	gems
Colombia	1984–	gold
Côte d'Ivoire	2002–	Diamonds
Democratic Republic of the Congo	1996–1997, 1998–	copper, coltan, diamonds, gold, cobalt, tin
Indonesia–West Papua	1969–	copper, gold
Liberia	1989–2003	diamonds, iron, gold
Papua New Guinea–Bougainville	1989–1998	copper, gold

Shannon O'Lear (2005) and Ross (2003) note that the presence of abundant resources with the following characteristics may provoke separatist rebellions: physically concentrated, extracted through capital-intensive processes that necessitate investment, and offering few benefits to local, unskilled workers, but more benefits to the state and large extraction firms. Many authors contend that the linkage between abundant mineral resources and intrastate conflict is the result of a broader phenomenon: the resource curse. Work by Richard Auty (2001), Alan Gelb (1988), and Jeffrey Sachs and Andrew Warner (2001), among others, found that mineral-exporting countries suffer disproportionately from poor economic performance, slow economic growth, high poverty rates, unequal income distributions, high corruption levels, and authoritarian governance. Scott Pegg (2003) focuses on three facets of the so-called curse: internationalization, centralization, and privatization. Internationalization refers to the dependence of the state on the sale of minerals in an external global economy for hard currency. Such dependence reduces the need for domestic taxation and state building. Paul Collier and Anke Hoeffler (2000, 26) found that "the extent of primary commodity exports is the largest single influence on the risk of conflict."

Centralization is the exclusive ownership of minerals by the state, with externally generated commodity revenues paid directly into the national treasury—a process that encourages corruption and cronyism. Privatization occurs when the state's leader relies on personal networks of control rather than formal bureaucracies. This combination of factors leads to a dependence on the commercial exploitation of resources. It also fosters violence as outside factions vie for access to revenues, and the state resorts to military, paramilitary, or mercenary power to protect those revenue-generating activities.

More generally, the literature suggests that many civil conflicts are about the appropriation of mineral rents. This is termed the greed motive for conflict: the opportunity for a nonstate group to capture the ruling group's resource rents (Collier and Hoeffler 2004). The allure of the greed motive has been exacerbated by the disengagement of superpowers willing to prop up governments with aid and a liberalized world economy willing to invest anywhere on the globe regardless of the on-the-ground situation. In response, sovereign and nonsovereign leaders have adopted market-oriented strategies to finance their governments, personal kleptocratic tendencies, or irredentist activities (Pegg 2003; Olsson and Fors 2004; Bannon and Collier 2003).

Ross (2004, 338) states that "civil war and resource dependence might be independently caused by some unmeasured third variable, such as the weak rule of law." Hence, countries with institutions that promote accountability and state competence will tend to benefit from the exploitation of mineral resources, because these institutions ameliorate the perverse political incentives that such resource wealth creates. Countries without functioning institutions are more likely to suffer from a resource curse (Tilton and Davis 2008; Robinson, Torvik, and Verdier 2006).

Limited access to resources due to their physical distribution (locational scarcity) can also be motivating reasons for intrastate disputes, particularly if other factors such as ethnic tensions are present (O'Lear 2005). Collier and Hoeffler (2004) point also to grievances as a motivation in civil wars. With respect to minerals, this would mean "deliberate institutional differences, installed by the ruling group, between formal and informal sector [mineral] production," including property rights and rule of law (Olsson and Fors 2004, 322). Locational scarcity leads to internal fights between rebel groups for control of the resource (as distinct from attempts to claim mineral-rich territory and secede). In particular, minerals whose extraction requires minimal technology, no involvement of the state machinery, and are valuable yet small (i.e., lootable) are often utilized to fund rebels, or groups fighting to overthrow governments. Conflict diamonds are the most well-known example (Global Witness 2006b; Ross 2003; Keen 1998).

The Interstate and International Nature of Mineral Conflicts

All the conflicts listed in table 10.1 are, or were, over access to minerals, or were funded by the sale of minerals. While they are internal conflicts, they are also interstate if the warring parties are sponsored by foreign governments—for example, militias sponsored by Uganda that have operated in the Democratic Republic of the Congo. Internal mineral conflicts also become interstate when foreign governments have direct legal, economic, or political roles in the conflicts, or in the country where the conflict is taking place (Ballentine and Nitzchke 2005). For instance, recently China has demonstrated a willingness to do business with African leaders (that governments in developed nations consider corrupt or unstable) in exchange for mineral concessions. Examples include but are not limited to Zimbabwe (platinum), the Democratic Republic of the Congo (copper, cobalt, and gold), and Zambia (copper).

An intrastate conflict inevitably becomes internationalized when the commodities involved are sold in international markets. If a profitable sale is ensured regardless of the circumstances surrounding the commodity's production, multinational mining firms that work in conflict areas and companies in other parts of the world that purchase these resources are consequently complicit in violence and conflict in other nations. In some cases the complicity is direct, such as when a firm pays either the state or a private military for protection, or submits to extortion from rebel groups within the country.

Pegg (2003) has created a typology of conflict that is based on a two-by-two extractor/security matrix. Extraction is undertaken by public or private entities, and security is likewise supplied by public or private groups. State mining firms guarded by the state's military might seem to be an internal affair, a situation with the potential for domestic human rights abuse, but to the degree that the mineral market is internationalized, any conflict will be as well. Private mining firms have depended on the state militaries to provide security, even to the point of flying troops in corporate planes or paying for the government's purchase of arms or weapons. For example, Freeport Indonesia PT (a U.S. firm) has paid the Indonesian military for protection of the Grasberg Mine in West Papua New Guinea (Brown 2005), and Aurora Gold (an Australian mining firm) has paid BRIMOB (the Indonesian government security forces) to provide protection at its mine in Borneo (Corporate Watch 2002). There has been violent conflict at both locations. Private mining firms often turn to private security forces for protection, and those forces have sometimes been willing to use violence to advance the firms' interests (Singer 2003). State mining firms also use private security. The governments of Angola and Sierra Leone, for instance, hired a private military company (Executive Outcomes, a South African company) to protect the state diamond mines. Again, there have been accusations of violent confrontations.

Social Sources of Conflict

Finally, researchers such as Jason Switzer (2001), Klare (2002), MMSD (2002), Ross (2004), and many others have pointed out that mineral extraction itself causes conflict stemming from community grievances over issues such as environmental impacts, community dislocation, the unfair treatment of workers, and physical violence. There are numerous examples (MMSD 2002). In the case of Newmont Mining in Sulawesi, Indonesia, the citizens living near the mine accused the company of polluting a local bay with arsenic and mercury through the disposal of tailings at sea via a pipeline.[5] Indonesian courts brought suit against both the company and the chief of the Newmont unit in Indonesia; the chief was jailed briefly and denied exit from the island for more than two years.

The development of gold mines in Ghana started in the early 1990s and has continued since that time, resulting in the displacement and resettlement of local people (Downing 2002). Displaced farmers were offered inadequate or no compensation for their lost livelihoods, private firms were used to remove prior residents from mine sites by force if necessary, and there has been evidence that known human rights abusers have been providing security at the mines (Robinson 2006). Then there is the case of the Panguan Mine in Bougainville, Papua New Guinea, where a dispute over the distribution of revenues (economic rent) led to a civil war, during which approximately ten thousand people died (Mack 2001).

As the preceding section illustrates, conflict over access to and control of minerals is unfortunately commonplace. The conflicts are driven by a desire to control the resource along with the revenue that it generates, the willingness to use force to do so, and disagreements over how mining is conducted. Hence, many conflicts are driven by locational, political, and social scarcity. In modern times, no interstate conflicts have been driven by depletion, yet this situation could change if the demand for commodities rises dramatically again. In response, efforts to define cooperative solutions have gained momentum in recent years.

Cooperation over Minerals

Cooperation with respect to minerals can take many forms, have a variety of drivers, and lead to many different types of agreements. Most often cooperative actions are focused on mineral supply activities: information gathering, production over the mine life cycle (i.e., exploration, mine development, extraction, and mine closure), stockpiling, and trade. Cooperation can take place within the private sector (i.e., investors, exploration and mine companies, and traders), among governments both externally and internally (i.e., country to country, or between national and local administrative entities), and with or within civil society (i.e., nongovernmental organizations [NGOs], interested groups, and individuals).

Cooperation has many aspects, but for the purpose of this chapter, we have chosen to define it in the following manner: the process of working or acting together, involving at least two parties, where everyone can gain, benefit, or at least not suffer loss. There are many characteristics of cooperation that may be applicable to natural resources. Particular to minerals, we acknowledge a number of aspects. First, cooperation can take place among states, regions, and companies as well as different combinations of these "players." The most common stakeholders are governments (local, national, and international), NGOs, companies, and in some cases the research community. Intergovernmental organizations can act as facilitators, overseers, or enforcers of cooperation. Second, after the cooperation is agreed on (established) and commences, events do not necessarily proceed as expected. Since the world is dynamic, circumstances (including political ones, and in the case of minerals, market conditions) may change, with mild or severe consequences for the cooperation and its outcomes. Third, the political framework (from the broad geopolitical to the smaller local scale) within which the cooperation takes place cannot, in most cases, be controlled by the cooperating parties. In the instance of minerals cooperation, outside factors can hamper a cooperative initiative. Fourth, cooperation strongly depends on the would-be cooperators' capacity (developed knowledge, skills, and resources) to achieve desired goals, which is why capacity-building activities are often a part of minerals cooperation. Fifth, there are situations where the cooperation is advanced by outside parties. Forced cooperation, however, cannot be expected to succeed, especially in cases where the participants do not directly see or experience any benefits. Finally, in some circumstances, cooperation begins voluntarily because the benefits of doing so are recognized. Over time, the cooperation may formalize with written agreements, and may eventually evolve into a mandatory requirement if it is legalized by governments or courts.

International cooperation in the minerals sphere is facilitated by various incentives. The following section begins with sustainable development and corporate social responsibility as a framework of cooperation, followed by a discussion of market-, policy-, and consensus-based incentives for cooperation. In addition, various cooperative mechanisms are described.

Drivers of Cooperation

Sustainable Development and Corporate Social Responsibility as Frameworks for Cooperation Incentives for cooperation are frequently motivated by the paradigms of sustainable development and corporate social responsibility. These notions are fairly recent phenomena and continue to frame much of the debate about cooperative activities, including those related to minerals.

Initially, discussions about the role of natural resources in sustainability tended to focus on the need to sustain ecosystems and maintain biodiversity.[6] For example, sustainable forest management requires that the capacity of forests to maintain their health, productivity, diversity, and overall integrity be protected, in the long run, in the context of human activity (USDA FS 2004). The fundamental goal is sustaining the ecosystem. Given the early rhetoric of sustainable development, it was thought inappropriate to speak of mineral resources as being sustainable in the same way as are ecosystems or biological resources given that they are nonrenewable. Furthermore, mining does negatively impact the environment. Together, these facts have led many people to express the simplistic view that mining is either inconsistent with sustainability (since once extracted, the resource is "gone"), anathema (as primarily a source of pollutants and environmental degradation), or of secondary importance (merely a source of virgin materials for which recycled materials or renewable resources can and should be substituted) (Shields, Šolar, and Langer 2006).

In reality, sustainable development involves managing resources in a way that is conducive to long-term wealth creation and the maintenance of capital (natural, social, human, economic, and physical). This perspective extends naturally to mineral resources, which are themselves a form of endowed, natural capital and are an important source of wealth creation. As a result, the discussion about minerals in sustainability now focuses on replacing depleted mineral capital with other forms of capital, environmental protection, and a fair and just distribution of the risks and benefits, as well as ensuring that the contribution of a mine is net positive over the life of the project, from exploration through postclosure (Shields, Šolar, and Miller 2007; MMSD NA 2002).

Although mineral resources were not addressed in Agenda 21 of the Earth Summit in 1992, they were included in the Johannesburg Plan of Implementation in 2002, in paragraph 46, which states in part,

> Enhancing the contribution of mining, minerals and metals to sustainable development includes actions at all levels to: (a) support efforts to address the environmental, economic, health, and social impacts and benefits of mining, minerals and metals throughout their life cycle, including worker's health and safety, and use a range of partnerships . . . to promote transparency and accountability for sustainable mining and minerals development. (UN 2002)

One way that firms are responding to the sustainability charge is by implementing the principles of corporate social responsibility (CSR), which is a form of business behavior that leads firms to voluntarily contribute to a better society and a cleaner environment (EC 2001). The goal of CSR is to raise standards of social development, environmental protection, and respect of fundamental rights by embracing open governance, reconciling the interests of stakeholders, and taking an overall approach to quality. CSR requires firm owners (shareholders) and management to act more responsibly toward employees in areas of working conditions, health, and safety, but also with regard to professional development (education), health insurance, and retirement funds. The rights and obligations of local communities, NGOs (mostly environmental), and other traditionally external stakeholders are recognized and respected.

Firms implement CSR and promote their CSR strategies for pragmatic reasons; doing so is good business. CSR can increase long-term business viability, including growth and profits. It sends a signal to stakeholders that the firm is a good and responsible corporate citizen (WBCSD 2000). And most important for the issue at hand, following sustainability principles and implementing CSR requires cooperation with host governments, local communities, workers, and other stakeholders, which in turn reduces the potential for conflict with them. States likewise encourage firms to practice CSR because they see the potential for reduced conflict in internationalized mining. At the G-8 Summit in Heiligendamm, Germany, June 6–8, 2007, participating nations adopted the "Responsibility for Raw Materials: Transparency and Sustainable Growth" declaration. The statement focused on the significance of the extractive sector and the benefits of CSR (G-8 2007).

Not all mining firms, whether state-owned or private, consider sustainability or CSR necessary or worthwhile activities. But firms that refuse to implement responsible business practices, that continue to extract and process minerals without regard for the welfare of their workers or the need to work collaboratively with nearby communities, are increasingly being pressured by labor unions, NGOs, and governments to change their corporate behavior.

Market-Based Incentives Markets play a central role in encouraging the exploration for, production, and distribution of minerals. Thus, it is worthwhile considering the degree to which cooperation is a factor in market functioning or market-based decisions. As we have noted before, the strong demand for metals (high prices) has led to an increased interest in minerals that reside on or beneath the ocean floor, in deep-seated, concealed deposits, and close to the surface, but in remote, inhospitable areas. This interest does not focus exclusively on the base and precious metals but also on those metals and industrial minerals that are vital components of strategically important, added-value products.

The high mineral prices in recent years have made some deposits profitable to produce and brought others to the brink of economic viability. This situation has created incentives for firms and governments to acquire commodity producers, sometimes in the form of "minority stakes in opaque companies in poorly governed countries that would otherwise make little business sense" (Rajan 2006, 54). The justification for such actions is the need for economic security. But Raghuram Rajan argues that these investments are not the best way for nations to hedge price risk. Rather, the supply should be ensured through world markets that are well informed, competitive, and function in a business environment that is transparent and predictable. Financial institutions support the creation of such a business environment by insisting that firms implement sustainable practices, including gaining prior consent from impacted communities for their presence and operations (World Bank 2003.) Financial institutions that adopt the Equator Principles of 2006 agree to only provide loans for projects that conform to nine principles that together are intended to ensure socially responsible development and sound environmental management (Secretariat for Equator Principles 2006).

Policy-Based Incentives Many existing mineral policies reflect the issues and interests of the prior eras when they were put in place. The public policies of the nineteenth and twentieth centuries embodied the societal interest in settlement, industrial development, and economic expansion. Current mining policies, including those rewritten in the 1990s, all address landownership, access, taxation, trade, and employment. Much less frequently do they address the transformation of resource capital into other forms of capital. Most existing policies have three characteristics that are relevant in this case. They tend to be stand-alone in nature, and lack a connection with or reference to other policies relevant to minerals management. They seldom address the full scope of concern described at the beginning of this chapter, although James Otto and John Cordes (2002) did note that as sustainable development is applied to mining, policymakers may focus more on intergenerational equity. And finally, few policies formally pay attention to the need for cooperation as a necessary part of minerals activities.

One way to encourage actions that will be cooperative and supportive of sustainable outcomes is to revise policies so as to incorporate the principles and goals of sustainability in a manner consistent with the social structure and desires of the nation. With respect to minerals, this means policies that: facilitate the transformation of natural mineral capital into built physical, economic, environmental, or social capital of equal or greater value; ensure that the environmental and social impacts of mining are minimized, and their costs are incorporated into production functions; address the benefit/risk allocations intra- and intergenerationally; and require transparency, information sharing, and collaboration/cooperation. It is also essential that a sustainable mineral policy be correlated and consistent with other governmental policies. Several countries have already employed such policies, including Canada (NRCan 1996), India, and the United Kingdom (UK 1999).[7]

Consensus-Based Incentives In the preceding section on conflict we discussed the prevalence of civil strife and violence in some developing nations that have significant resource endowments. It was suggested that such internal conflicts were likewise interstate or internationalized in the sense that foreign governments, multinationals, and private companies were directly or indirectly involved. International cooperation will be required to avoid and end such internal conflicts. In fact, in recent years there has been both a global call for actions to avoid civil wars and widening agreement on the steps that need to be taken to prevent natural resource–driven conflicts. Such a broad international consensus provides both a political incentive for nations to follow these steps and a public incentive for firms that do not want to be seen as acting outside agreed-on norms of behavior to do so as well. These steps include: strengthening the governance of the international trade in mineral commodities; halting the financing of illicit commodities, including criminalizing the sale of booty futures;[8] promoting responsible behavior by large and small companies, and attracting reputable companies to risky environments; monitoring and assessing developments, and publishing the results of assessments; fighting poverty and creating sustainable livelihoods, in communities as well as for artisanal and small-scale miners; raising economic growth rates in host nations; reducing macroeconomic dependence on minerals and vulnerability to mineral market fluctuations; increasing participation, dialogue, and partnership; empowering local communities through information access; addressing gaps in policy and governance; making payments from multinational firms to governments transparent; and engaging in preventive diplomacy.

A number of multistakeholder collaborative initiatives have been created in recent years that address the various aspects of international consensus outlined above. These include:

- The Extractive Industries Transparency Initiative (EITI) is a coalition of governments, companies, civil society groups, investors, and international organizations. The EITI sets a global standard for companies to publish what they pay and for governments to disclose what they receive.[9]
- The Publish What You Pay coalition of over 280 NGOs worldwide calls for the mandatory disclosure of the payments made by oil, gas, and mining companies to all governments for the extraction of natural resources.[10]
- The Kimberley Process aims to regulate diamond mining and trade through a certifying and tracking scheme to preserve the legitimate diamond trade while excluding conflict diamonds from international markets.[11]
- Global Witness was the first organization that sought to break the links between the exploitation of natural resources, and conflict and corruption, working to expose the connection between natural resource exploitation and human rights abuses, including campaigns on oil and diamonds.[12]

Characterizing Cooperation: Between Whom and in What Manner

Cooperation takes many forms and occurs between a wide variety of stakeholders. In this section, we present different types of cooperative actions enumerated by the parties involved. The discussion below is intended not to be exhaustive but rather to illustrate the breadth of scope for cooperation and provide a limited set of specific examples. In each case, cooperation has taken place because of concerns about real or perceived mineral scarcity, and/or a desire to prevent or end interstate conflict. The various cooperative incentives and pillars mentioned above are implicit in the discussion.

Between Producing and Consuming Nations Cooperation between producing and consuming nations can take the form of free trade agreements. For example, the governments of Australia and Japan commenced negotiations on the Free Trade Agreement (FTA) in April 2007, following the conclusion of a joint government study on the feasibility of a bilateral FTA. The study concluded that a comprehensive and World Trade Organization–consistent FTA between Australia and Japan would bring significant benefits to both countries (AU 2008). Minerals trade is one reason for the interest in the FTA. Japan relies on Australia as an important supplier of minerals commodities, and is likely to continue to do so for the medium and long term (Minerals Council of Australia 2005). The intensity of this relationship is mutual; Japan absorbs a significant proportion of Australia's minerals exports, though not as much as China does.

Similarly, in 1988 China and Canada signed the first Memorandum of Understanding (MOU) on cooperation in minerals and metals. In 2000, an MOU on cooperation in the sustainable development of minerals and metals was signed as well between the two countries.[13] Between 2000 and 2004, China streamlined exploration licensing and permitting regulations and procedures, which became more transparent. Mining taxation incentives also became more favorable for foreign investment in China (Nash 2006; NRCan 2006). The MOUs and economic incentives have encouraged Canadian firms to conduct exploration activities and invest in mines in China.

Germany has developed a mineral policy that reflects its market positions. There was no metal mining in Germany in 2007; the country is 100 percent dependant on imports sourced from all over the world. The policy has three pillars: the economic viability of the German economy; sustainable development; and securing raw materials supply, which includes building cooperative links with the developing nations where the minerals are produced to ensure continued access to resources. The German government states that it has no subsidies, no national stockpile, and no market-distorting measures. Its policy promotes the use of political instruments to overcome trade barriers and provide the ground for free trade (Homberg 2007).

Cooperation can also extend to information gathering and dissemination, and the assessment of mineral resources. We noted in the discussion of scarcity that there is a need to discover new mineral deposits to replace those being depleted. As Klaus Schultz and Joseph Briskey (2003, 1) wrote,

> The growing demand for mineral resources requires continued exploration and development of as-yet-undiscovered mineral deposits. Informed planning and decisions concerning sustainable resource development require a long-term global perspective and an integrated approach to land-use, resource, and envi-

ronmental management. This integrated approach, in turn, requires that unbiased information be available on the global distribution of identified and especially undiscovered mineral resources, the economic factors influencing their development, and the environmental consequences of their exploitation.

In response, the U.S. Geological Survey is conducting the Global Mineral Resource Assessment Project, a cooperative international effort to assess the world's undiscovered nonfuel mineral resources.[14] The participating nations include, but are not limited to, Chile, China, France, Georgia, South Korea, Mongolia, Russia, and Turkey.

Among Consuming Nations The nations of the European Union (EU) have been addressing the issue of mineral availability in recent years because of concerns about the potential for inadequate supply to member state economies. On May 21, 2007, the Council of Ministers (of the High Level Group on Competitiveness) requested that the European Commission work to

> develop a coherent political approach with regard to raw materials supplies for industry, including all relevant areas of policy (foreign affairs, trade, environmental, development and research and innovation policy) and to identify appropriate measures for cost-effective, reliable and environmentally friendly access to and exploitation of natural resources, secondary raw materials and recyclable waste, especially concerning third-country markets. (EC 2007, 2)

This is noteworthy in part because to date member nations have had their own minerals policies, but the EU has not had a policy that applies to all member states. Clearly, concerns about scarcity have led to a reconsideration of that position.

The main focus of the raw materials policy is on the nonenergy extractive industry. In the course of preliminary work, three main challenges have been identified. Dealing with them will require cooperation first among EU nations, and subsequently between the EU and the mineral-exporting nations. They include access to raw materials on world markets under undistorted conditions, a sustainable supply of raw materials from European sources, and an increase of resource efficiency and the promotion of recycling (EC 2008).

Among Producing Nations The purpose of the Association of Southeast Asian Nations (ASEAN) is to accelerate economic growth, social progress, and cultural development in member nations. Given the importance of minerals to economic growth, such as industrial and construction inputs as well as export commodities, the ASEAN countries have developed the

ASEAN Minerals Cooperation Action Plan 2005–2010 (ASEAN 2005). Its purpose is to create a profitable and growing ASEAN minerals sector by enhancing trade and investment while strengthening cooperation and the capacity for sustainable mineral development in the region. The strategies to be implemented in the plan include "facilitating and enhancing trade and investment in minerals, promoting environmentally and socially sustainable mineral development, and strengthening institutional and human capacities in the ASEAN minerals sector" (ASEAN 2005, 3).

The Mines Ministries of the Americas (CAMMA) is a Western Hemispheric organization that includes twenty-three countries in the Americas.[15] It has goals similar to those in the ASEAN plan. The Santiago Declaration recognized that mining contributed decisively to the economies of mining countries and reinforced the importance of hemispheric cooperation (CAMMA 1996). CAMMA held formal ministerial meetings and workshops through 2005. There has been agreement on the need to strengthen the network of existing partnerships across the hemisphere in order to collectively address the challenges facing minerals and metals development as well as benefit from the sharing of information and best practices to foster the development of common approaches and policies for the sustainable development of minerals and metals. In the Argentine Republic Declaration of 2005, member nations agreed to continue work on the main themes of sustainable development and market access, maintain stakeholder dialogues, and contribute to the United Nations Commission on Sustainable Development policy review of mining in 2010, among other activities (CAMMA 2005).

The Partnership Africa-Canada (PAC 2005) provides another example of producing country interstate cooperation. The partnership works to foster research on and implementation of policies that contribute to sustainable development in Africa. One of the initiative's focuses is conflict diamonds. Canada and several African countries produce diamonds, and the need for cooperation emerged partly in reaction to the problem of trade in conflict diamonds. The partnership comprises policy research, education, publication, and advocacy to ensure that the international diamond industry operates legally, openly, and for the primary benefit of the countries where diamonds originate, becoming an asset for—rather than a detriment to—peaceful, long-term development. The Partnership Africa-Canada is an active participant in the Diamond Development Initiative, a process addressing the artisanal diamond-mining sector, and the Kimberley Process. Canada is not affected economically by the production of conflict diamonds but instead participates in this cooperation as part of a broader engagement in Africa based on diplomatic and altruistic motivations.

Through Intergovernmental Bodies Countries have been actively cooperating in a variety of intergovernmental bodies for many years. Several of these have been described in the preceding sections; looking at all existing cooperation is far beyond the scope of this chapter. Instead we focus on three examples, one of which is an ongoing process that is an outgrowth of the 2002 World Summit on Sustainable Development language on minerals and mining. The second one, UNCLOS, predates even the Earth Summit, but has recently become the focus of renewed attention. Finally, we briefly consider the United Nations Conference on Trade and Development's Integrated Programme for Commodities.

Following the conclusion of the World Summit on Sustainable Development, the governments of Canada and South Africa launched a partnership activity titled the Intergovernmental Forum on Mining, Minerals, Metals, and Sustainable Development (IFMMMSD 2005). Their goal was to bring together governments with an interest in these topics to work to fulfill the priorities for mining that were stated in paragraph 46 of the Johannesburg Plan of Implementation through enhanced capacity for governance (UN 2002). The thirty-eight delegations to the forum comprise government, industry, and NGO representatives. Nonmember governments as well as multilateral and international agencies have observer status. To date the forum has addressed: the national mining policy framework considering priorities, data collection, and management; a template for a mineral policy framework; a framework on investors' perceptions of country risks; country perceptions of investor risks; environmental impact assessments; policies that best assure an equitable local, regional, and national distribution of benefits; financial surety and environmental protection including mine rehabilitation and closure; generating benefits from mining investments; and communities and mining (Bourassa 2008).

In 2010, the United Nations Conference on Sustainable Development (UNCSD 18) conducted a review of progress made toward the achievement of minerals-related goals laid out in paragraph 46 of the Plan of Implementation from WSSD in 2002. During the 2011 cycle (UNCSD 19), policy proposals for international minerals management will be debated and voted on.[16] Using these year dates as benchmarks, the forum is also proposing to create a joint mineral policy statement that would cover the entire mineral sector life cycle, address all sector issues, and list agreed-on policy elements.

International minerals cooperation is likewise evidenced through UNCLOS. The convention defines the rights and responsibilities of nations in their use of the world's oceans, establishing guidelines for businesses, the environment, and the management of marine natural resources. It treats waters more than 200 nautical miles off coasts as under the purview of a new international organization, the International Seabed Authority, unless certain geologic parameters of contiguity exist. The International Seabed Authority has the right to set production controls for ocean mining, drilling, and fishing, regulate exploration, and issue permits, all within its own tribunal. The tribunal, through its Seabed Disputes Chamber, has the exclusive jurisdiction to settle conflicts arising from the implementation of the regime for mining minerals from the deep seabed (Nandan 2006).

UNCLOS came into force in 1994, after having gained the requisite sixty signatories. From its inception it has been controversial, particularly with developed nations, many of whom believed that the proposed new international legal regime would limit their national sovereignty. Concerns were also expressed that seabed mineral rights would be allocated by the UNCLOS tribunal rather than being claimed by nations. Currently, 155 states are party to UNCLOS.

Both Russia and Canada are signatories to the treaty (whereas the United States is not), and they are now using UNCLOS as a forum within which to adjudicate competing claims for the Arctic seabed. On August 2, 2007, a Russian submarine crew planted a Russian flag at the North Pole. The Russian government is claiming the territory, and the vast energy and mineral resources residing there, based on the assertion that the Lomonosov Ridge is an integral part of its country's landmass, as defined under UNCLOS. Natural Resources Canada and Fisheries and Oceans Canada are conducting Arctic surveys to acquire the scientific data needed to substantiate Canada's submission to the United Nations Commission on the Limits of the Continental Shelf, claiming that its sovereign rights extend beyond 200 nautical miles (Bridges 2008; Lunn 2008). Adjudication of this matter within the Seabed Disputes Chamber may prevent the two nations from resorting to force in order to protect their interests.

The United Nations Conference on Trade and Development focuses on the rules of international trade and started the Integrated Program for Commodities.[17] This program stipulated that agreements for eighteen specified commodities would be negotiated or renegotiated with the principal aim of avoiding excessive price fluctuations, and stabilizing commodity prices at levels remunerative to the producers and equitable to consumers. The International Copper Study Group, an outgrowth of the Integrated Program for Commodities, is an intergovernmental body of seventeen countries and the EU, and provides a forum within which industry, its associations, and governments can meet to discuss common problems and objectives with respect to copper. The group has three major objectives: promoting international cooperation, providing a global forum, and increasing market transparency.[18]

Among Firms Cooperation among mining firms historically has taken two forms: property shares and voluntary industry associations. Firms (whether private or government owned) gain access to minerals by purchasing a percentage of a deposit or operation owned by another public or private entity. Such purchases are driven by a variety of business motives. With respect to scarcity and cooperation, these can include the need to replace depleted reserves, a desire to have access to minerals that are the property of another firm, or the need to simply avoid conflict with a host government-owned or privately held mining firm.

Mining firms also cooperate through industry associations. Participants voluntarily enter into an agreement to accomplish common goals. Associations are based on a purpose, and have a clear vision, mission, regulations, and organizational structure. Mining associations are founded and funded by corporations that operate in a specific mining sector. In most cases they promote the industry (through advertising, education, and lobbying), but they can also encourage collaboration between companies in many areas, such as standardization, and responsible corporate behavior through the development of association policies that members must, or are encouraged to, adopt.

Mining organizations in the United States include the National Mining Association, the National Stone Sand and Gravel Association, and the Industrial Minerals Association–North America. All have policies and position statements on sustainable development and best practices. Associations active on the EU level cover all major types of minerals resources: metals (Euromines), industrial minerals (IMA-Europe), and UEPG (aggregates). They are active in promoting their interests through lobbying as well as by attending meetings and issuing position papers. All these organizations encourage member firms to practice sustainable development.

Another group, with a different purpose than traditional industry associations, is the International Council on Mining and Metals (ICMM 2008). This organization of mining firms is "a platform for industry and other key stakeholders to share challenges and develop solutions based on sound science and the principles of sustainable development."[19] The ICMM has published numerous reports on best practices in areas such as materials stewardship, environmental management, safety, corporate citizenship, and working with indigenous peoples. The Global Reporting Initiative (GRI)–ICMM Working Group wrote the Mining and Metals Sector Supplement to the GRI 2002 Guidelines.[20] The GRI develops and disseminates globally applicable "Sustainability Reporting Guidelines" for voluntary use by organizations reporting on the economic, social, and environmental sustainability aspects of their operations.[21] In addition, the ICMM has participated actively in the Extractive Industries Review put out by the World Bank (2003), and endorsed the Review's final focus on sustainable development and poverty alleviation.

Between Firms and Communities Multinational mining firms are increasingly looking for ways to work cooperatively with communities in the regions where they are developing and operating mines. They are interested in doing so because community opposition presents risks to project financing, construction, and operation as well as to the firm's reputation and its working relationship with host governments (Ethical-Funds 2008). Most of all, local or regional opposition to mining can lead to conflict.

Barrick Argentina provides an example of how mining firms can build cooperative relationships. The firm worked with the local communities near its Veladero Mine to create and implement the Sustainable and Community Development Plan (Claudeville 2008). The plan has four parts: health and safety, education and training, economic growth, and community relations. The firm has taken the following actions: strengthening infant and juvenile health and dental programs; providing hygiene and safety as well as defensive driving training; helping local farmers to create the "El Porvenir" cooperative, through which they market their bean crop directly to retailers rather than selling to wholesale buyers; and setting up a cooperative monitoring program with community representatives. This latter action is particularly significant because community participation in measuring environmental impacts increases trust and reduces opposition to mining.

Conclusion

This book considers interstate cooperation over scarce resources. To investigate how this topic applies to minerals, we posed a series of questions: Are minerals scarce? Is there interstate conflict over minerals, and if so, is scarcity the cause? Is there interstate cooperation with respect to minerals, and if so, is scarcity the motivation? We will first review the answers to these questions and then consider what they mean when taken as a whole.

The specter of mineral resource scarcity has arisen repeatedly, most recently in the past eight years in response to increased demand from China and India. As our market response model illustrated, an imbalance between supply and demand can lead to price increases, which was the case until late in 2008. The recent price increase might have been a reflection of concern about reduced commodity stocks; alternatively, it might have been indicative of inadequate flow through the mineral market system. A review of the existing literature led us to the conclusion that the fixed stock perspective is not applicable to the vast majority of minerals; the physical depletion of stocks is not a serious concern. Conversely, the opportunity cost perspective argues that the higher the price, the more markets will seek new commodity sources through substitution and recycling. Resources will thus never be physically depleted, but instead will be economically depleted. A further consideration of the economic measures of scarcity led to the conclusion that price, extraction cost, and user cost are all imperfect measures of true physical scarcity.

We then turned to alternative conceptions of scarcity, all of which are based on inadequate flow through the minerals market system rather than physical depletion. Situational scarcity is caused by a lack of human, physical, or financial capital. The recent increase in demand from Asia and South Asia highlighted the need for increased investment in minerals-related capital. Its subset, locational scarcity, occurs when commodities are known, but do not exist in the same geographic region as where they are needed. Political scarcity, which is driven by the actions and choices of governments, has been receiving considerable attention in recent years, and is linked to both national and economic security concerns. We also identified social scarcity as being of increasing importance because of the broadened scope of concern regarding the mineral supply. Our review of scarcity leads us to the view that while minerals are not physically scarce, situational, political, and social scarcity does occur.

We next turned to the issue of conflict over minerals, which is unfortunately both widespread and commonplace. We investigated the degree to which these conflicts are interstate and driven by scarcity. Many conflicts that are now seen as minerals related actually began for other reasons. Yet once the warring factions realized that the sale of minerals could be used to purchase arms or pay troops, the conflict expanded to include control of those resources. Our review of the literature on civil conflicts initiated over minerals found that both abundance and scarcity can be causative factors. Abundance can lead to the internationalization of commodity sales, and in the absence of good governance, can also result in the centralization of state ownership and the control of revenues, and the use of personal networks of control rather than state building. Together, these circumstances foster violence when those outside the power structure vie for access to revenues. Conversely, the uneven distribution of lootable resources (or locational scarcity) can also lead to conflict if the economic rent generated by mineral production is distributed only to a subset of a country's population, or if rebel groups vie for control and thus access to the rent.

While we found no recent cases of direct state-on-state conflict over minerals, we did identify instances where countries meddle in the civil conflicts of their neighbors. We have also argued that when countries or their state-owned mining firms do business with corrupt governments in countries experiencing civil strife, that conflict indirectly becomes interstate. More broadly, mineral conflicts become internationalized when the participants have different countries of origin. We identify both political and social scarcity drivers for such conflicts. Our review of conflict led us to conclude that fears about the consequences of situational, political, and social scarcity have the potential to trigger interstate conflict, but have not done so in recent years.

That cooperation, rather than conflict, is the preferable approach to dealing with scarcity is self-evident. The alternatives of war over resources, violent civil conflict, environmental degradation, or deprivation are unacceptable. Most recent forms of interstate or international cooperation can be framed within the context of sustainable development. Further, we have shown that there are three underlying drivers of cooperation: market-, policy-, and consensus-based incentives. We also reviewed a wide range of alternative forms of cooperation between a variety of parties. Virtually all arose from concerns about perceived or potential scarcity, the need to address the full scope of concern with respect to mineral supply, and a desire to prevent interstate or internationalized mineral conflicts.

Which type of cooperative framework is most appropriate for a specific situation will depend on the unique, physical characteristics of the mineral itself, the structure of the markets within which it is traded, and the nature of the societies that are party to production and trade. Openness to cooperation will depend on a country's or firm's history and culture. Policy, be it corporate, national, or intergovernmental, will be key since it is the backbone of cooperation. Transparent processes for creating sustainable mineral policies, and the collaborative implementation of those policies, will be essential as well. The purpose of many of the cooperative ventures we have described is to facilitate the development of policies that will increase the quality of governance, promote equity, regulate and ensure stable mineral market transactions, guarantee revenue flows, and encourage the practice of CSR. Interstate and international cooperation undertaken in this manner has the potential to prevent interstate conflict over mineral scarcity.

Notes

1. The Great Game refers to the tussle among Britain, Russia, and other countries for influence in the Middle East and Central Asia during the nineteenth century. The presence of oil was one of the drivers of the Great Game.

2. A full text of the convention is available at <http://www.un.org/Depts/los/convention_agreements/convention_overview_convention.htm> (accessed May 4, 2008).

3. Market disruptions resulting in price decreases also occur, but they have different drivers, such as economic recession (i.e., decreased demand), the discovery of new deposits, and releases from stockpiles or dumping, rather than scarcity.

4. The NRC committee's differing assessment of indium from that of the National Renewable Energy Laboratory is another example of how the use of different paradigms will lead to inconsistent availability assessments.

5. Donald Greenlees, "Indonesian court acquits Newmont Mining." *New York Times*, April 25, 2007. Available at <http://www.nytimes.com/2007/04/25/world/asia/25indo.html> (accessed 20 August 2008).

6. Sustainable development strives to improve the economy, environment, and quality of life for the current generation without compromising the ability of future generations to meet their needs. The foundation is the reconciliation of a society's development goals with the earth's environmental limits. The world's nations have agreed on the need to transition to a sustainable development path, first at the Earth Summit held in Rio de Janeiro, Brazil, in 1992 (UN 1992) and again in 2002 at the World Summit on Sustainable Development held in Johannesburg, South Africa (UN 2003).

7. Additional information is available at <http://mines.nic.in/> (accessed April 4, 2008).

8. Booty futures are the sale of rights to commodities that rebel groups do not yet control, but expect to if their rebellion is successful.

9. Additional information is available at <http://www.eitransparency.org/node> (accessed April 8, 2008).

10. Additional information is available at <http://www.publishwhatyoupay.org> (accessed April 8, 2008).

11. Additional information is available at <http://www.kimberleyprocess.com> (accessed April 8, 2008).

12. Additional information is available at <http://www.globalwitness.org/index.php> (accessed April 8, 2008).

13. The full text of the MOU is available at <http://www.nrcan-rncan.gc.ca/mms-smm/poli-poli/int-int/china-eng.htm> (accessed January 3, 2010).

14. Additional information is available at <http://minerals.usgs.gov/east/global/index.html> (accessed April 4, 2008).

15. Additional information is available at <http://www.camma.org> (accessed April 4, 2008).

16. The other United Nations Conference on Sustainable Development themes for 2010–2011 are transport, chemicals, waste management, and sustainable production and consumption.

17. Additional information is available at <http://www.unctad.info/en/Special-Unit-on-Commodities/> (accessed April 4, 2008).

18. Additional information is available at <http://www.icsg.org> (accessed October 7, 2008).

19. Additional information is available at <http://www.icmm.com> (accessed May 7, 2008).

20. The text of the Draft Final Mining and Metals Sector Supplement is at <http://www.globalreporting.org/ReportingFramework/SectorSupplements/MiningAndMetals/MiningAndMetals.htm> (accessed January 3, 2010).

21. Additional information is available at <http://www.globalreporting.org> (accessed January 3, 2010).

References

AID (U.S. Agency for International Development). 2004. *Minerals and Conflict: A Toolkit for Intervention*. Available at <http://www.usaid.gov/our_work/cross-cutting_programs/conflict/publications/docs/CMM_Minerals_and_Conflict_2004.pdf> (accessed April 4, 2008).

Alonso, Elisa, Frank Field, Jeremy Gregory, and Randolph Kirchain. 2007. "Material availability and the supply chain: Risks, effects, and responses." *Environmental Science and Technology* 41 (19): 6649–6656.

Andrews, Craig. 2008. Opening remarks at the World Mines Ministries Forum. February 29. Available at <http://www.wmmf.org/downloads.html> (accessed April 4, 2008).

ASEAN (Association of Southeast Asian Nations). 2005. *ASEAN Minerals Cooperation Action Plan 2005–2010*. Available at <http://www.aseansec.org/17706.pdf> (accessed April 4, 2008).

AU (Australia, Department of Foreign Affairs and Trade). 2008. *Australia-Japan Free Trade Agreement Negotiations*. Available at <http://www.dfat.gov.au/geo/japan/fta/index.html> (accessed April 4, 2008).

Auty, Richard, ed. 2001. *Resource Abundance and Economic Development*. Oxford: Oxford University Press.

Ayers, Robert. 2001. "Resources, scarcity, growth, and the environment." Center for the Management of Environmental Resources. Available at <http://ec.europa.eu/environment/enveco/waste/pdf/ayres.pdf> (accessed April 4, 2008).

Ballentine, Karen, and Heiko Nitzchke, eds. 2005. *Profiting from Peace: Managing the Resource Dimension of Civil War*. Boulder, CO: Lynne Rienner Publishers.

Ballentine, Karen, and Jake Sherman, eds. 2003. *The Political Economy of Armed Conflict: Beyond Greed and Grievance*. Boulder, CO: Lynne Rienner.

Bannon, Ian, and Paul Collier, eds. 2003. *Natural Resources and Violent Conflict: Options and Actions*. Washington, DC: World Bank.

Barnett, Harold, and Chandler Morse. 1963. *Scarcity and Growth*. Baltimore: Johns Hopkins University Press.

Baumol, William J., and Sue Anne Batey Blackman. 2007. "Natural resources." In *The Concise Encyclopedia of Economics*, ed. David R. Henderson. Available at <http://www.econlib.org/library/Enc/NaturalResources.html> (accessed April 4, 2008).

Bourassa, Andre. 2008. "Intergovernmental Forum on Mining, Minerals, Metals, and Sustainable Development." PowerPoint presentation at the World Mines Ministries Forum, Toronto, February 28–March 2. Available at <http://www.wmmf.org/downloads.html> (accessed April 4, 2008).

Bridges, Holly. 2008. "Canadian forces assist with a survey of the Arctic seabed." *Maple Leaf* (National Defense Canada), May 7. Available at <http://www.forces.gc.ca/site/Commun/ml-fe/vol_11/vol11_17/1117_full.pdf> (accessed January 7, 2009).

Brown, Oli. 2005. "Wealth for the few, poverty for the many: The resource curse-examples of poor governance/corporate mismanagement wasting natural resource wealth." Human Development Report Office Occasional Paper. Available at <http://hdr.undp.org/en/reports/global/hdr2005/papers/hdr2005_brown_oli_30.pdf> (accessed April 4, 2008).

CAMMA. (Annual Conference of the Mines Ministries of the Americas). 1996. Santiago Declaration. Available at <http://www.camma.org/declarations.asp?lang=en> (accessed May 12, 2010).

CAMMA. (Annual Conference of the Mines Ministries of the Americas). 2005. Argentine Republic Declaration. Available at <http://www.camma.org/docs/Decl%20ARG-Ver-0504%20-%20FIN%20AL.pdf> (accessed April 4, 2008).

Claudeville, Julio. 2008. "Barrick Argentina: Sustainable development and community relations." PowerPoint presentation at the International Mineral Processing Congress, Beijing, September 24.

Collier, Paul, and Anke Hoeffler. 2000. "Greed and grievance in civil war." Center for the Study of African Economies Working Paper CSAE WPS/2002–01. Available at <http://www.csae.ox.ac.uk/workingpapers/pdfs/2002-01text.pdf> (accessed April 4, 2008).

Collier, Paul, and Anke Hoeffler. 2004. "Greed and grievance in civil war." *Oxford Economic Papers* 56 (4): 563–595.

Corporate Watch. 2002. "Indonesia: Man shot at Australian gold mine." Available at <http://www.corpwatch.org/article.php?id=1388> (accessed April 27, 2008).

Darmstadter, Joel. 2001. "Summary of the interdisciplinary workshop on long-run availability of minerals: Geology, environment, economics." Workshop sponsored jointly by Resources for the Future and the Mining Minerals and Sustainable Development Project. Washington DC, April 22–23. Available at <http://www.rff.org/rff/Documents/RFF-Event-April01-Darmstadter-Sum.pdf> (accessed April 4, 2008).

Downing, Theodore. 2002. "Avoiding new poverty: Mining-induced displacement and resettlement." Mining Minerals and Sustainable Development Report 58. Available at <http://www.iied.org/pubs/pdfs/G00549.pdf> (accessed April 4, 2008).

EC (European Commission). 2001. *Promoting a European Framework for Corporate Social Responsibility*. Available at <http://eur-lex.europa.eu/LexUriServ/site/en/com/2001/com2001_0366en01.pdf> (accessed December 31, 2009).

EC (European Commission). 2007. "Public consultation on commission raw materials initiative." Background Paper. Available at <http://www.ec.europa.eu/enterprise/newsroom/cf/document.cfm?action=display&doc_id=633&user service_id=1> (accessed May 7, 2008).

EC (European Commission). 2008. "The raw materials initiative: Meeting our critical needs for growth and jobs in Europe." *Communication from the Commission to the European Parliament and the Council* 699. Available at <http://ec.europa.eu/enterprise/newsroom/cf/document.cfm?action=display&doc_id=894&userservice_id=1> (accessed January 3, 2010).

Ethical Funds. 2008. *Sustainability Perspectives—Winning the Social License to Operate: Resource Extraction with Free, Prior, and Informed Community Consent*. Available at <https://www.ethicalfunds.com/SiteCollectionDocuments/docs/FPIC.pdf> (accessed April 4, 2008).

G-8. 2007. *Growth and Responsibility in the World Economy: Summit Declaration*. June 7. Available at <http://www.state.gov/documents/organization/92264.pdf> (accessed May 7, 2008).

Gelb, Alan. 1988. *Oil Windfalls: Blessing or Curse?* New York: Oxford University Press.

GFMS (Gold Fields Mineral Services). 2008. "Spot metal prices and metal index". Available at <http://www.gfms.co.uk> (accessed April 4, 2009).

Global Witness. 2006a. "Natural resources in conflict." Available at <http://www.globalwitness.org/pages/en/natural_resources_in_conflict.html> (accessed April 4, 2008).

Global Witness. 2006b. *The Sinews of War: Eliminating the Trade in Conflict Resources*. Available at <http://www.globalwitness.org/media_library_detail.php/480/en/the_sinews_of_war> (accessed April 4, 2008).

Goeller, Harold E. 1972. "The ultimate mineral resource situation: An optimistic view." *Proceedings of the National Academy of Sciences of the United States of America* 69 (10): 2991–2992.

Gordon, Robert E., Bertram McInnes, and Thomas E. Graedel. 2006. "Metal stocks and sustainability." *Proceedings of the National Academy of Sciences of the United States of America* 103 (5): 1209–1214.

Gordon, Robert, Bertram McInnes, and Thomas E. Graedel. 2007. "On the sustainability of metal supplies: A response to Tilton and Lagos." *Resources Policy* 32 (1–2): 24–28.

HIIK (Heidelberg Institute on International Conflict Research). 2004. "Conflict barometer 2004." Available at <http://www.hiik.de/konfliktbarometer/pdf/ConflictBarometer_2004.pdf> (accessed April 4, 2008).

Homberg, Doris. 2007. "Minerals information significant to the German raw materials policy." PowerPoint presentation at the Raw Materials Initiative Workshop INFRA 25708. Ljubljana, December 10–11. Available at <http://www.geo-zs.si/UserFiles/677/File/TAIEX/24_Doris%20Homberg.pdf> (accessed April 4, 2008).

ICMM (International Council on Mining and Metals). 2008. Available at <http://www.icmm.com> (accessed May 4, 2008).

IFMMMSD (Intergovernmental Forum on Mining, Minerals, Metals, and Sustainable Development). 2005. "Introduction." Available at <http://www.globaldialogue.info> (accessed April 4, 2008).

Keen, David. 1998. "The economic function of violence in civil wars." Adelphi Paper 320. Oxford: Oxford University Press.

Klare, Michael T. 2002. *Resources Wars: The New Landscape of Global Conflict*. New York: Henry Holt.

Krautkraemer, Jeffrey. 2005. "Economics of natural resource scarcity: The state of the debate." Resources for the Future Discussion Paper 05-14. Washington, DC: Resources for the Future.

Le Billon, Philippe. 2003. "Getting it done: Instruments of enforcement." In *Natural Resources and Violent Conflict: Options and Actions*, ed. Ian Bannon and Paul Collier, 215–286. Washington, DC: World Bank.

LME (London Metal Exchange). 2008. "Non-ferrous metals cash buyer prices." Available at <http://www.lme.co.uk/non-ferrous.asp> (accessed April 20, 2009).

Lunn, Gary. 2008. Notes for a speech by the Honorable Gary Lunn, Canadian minister of natural resources, to the World Mines Ministries Forum, Toronto, March 2. Available at <http://www.nrcan-rncan.gc.ca/media/spedis/2008/200814-eng.php> (accessed May 4, 2008).

Mack, Andrew. 2001. "The private sector and conflict." UN Global Compact. Available at <http://www.unglobalcompact.org/issues/conflict_prevention/meetings_and_workshops/privateSector.html> (accessed October 10, 2008).

Mandel, Robert. 1988. *Conflict over the World's Resources*. New York: Greenwood.

McKelvey, Vincent E. 1972. "Mineral resource estimates and public policy." *American Scientist* 60: 32–40.

Meadows, Dennis L., Jørgen Randers, and Donella H. Meadows. 2004. *Limits to Growth: The 30 Year Update*. White River Junction, VT: Chelsea Green.

Meadows, Donella H., Dennis L. Meadows, Jørgen Randers, and William W. Behrens. 1972. *The Limits to Growth*. New York: Universe Books.

Menzie, David W. 2006. "Testimony before the Committee on Resources, Subcommittee on Energy and Mineral Resources, United States House of Representatives." Hearing on Energy and Mineral Requirements for Development of Renewable and Alternative Fuels Used for Transportation and Other Purposes. May 18. Available at <http://www.doi.gov/ocl/2006/RenewableAndAlternativeFuels.html> (accessed April 4, 2008).

Menzie, David W., Donald A. Singer, and John H. DeYoung, Jr. 2005. "Mineral resources and consumption in the twenty-first century." In *Scarcity and Growth Revisited*, ed. David R. Simpson, Michael A. Toman, and Robert U. Ayers, 33–53. Washington, DC: Resources for the Future.

Minerals Council of Australia. 2005. *An Australia-Japan Free Trade Agreement: The Minerals Industry Case*. Available at <http://www.minerals.org.au/__data/assets/pdf_file/0014/10391/MCAJapaneseFTA_SubmissionOct05.pdf> (accessed April 4, 2008).

MMSD (Mining, Minerals, and Sustainable Development). 2002. *Breaking New Ground: Mining, Minerals, and Sustainable Development (Final Report)*. Available at <http://www.iied.org/mmsd/finalreport> (accessed April 4, 2008).

MMSD NA (Mining, Minerals, and Sustainable Development—North America). 2002. *Seven Questions to Sustainability: How to Assess the Contribution of Mining and Minerals Activities*. Task 2 Work Group, MMSD—North America. Available at <http://www.iisd.org/pdf/2002/mmsd_sevenquestions.pdf> (accessed May 4, 2008).

Moll, Stephan, Stefan Bringezu, and Helmut Schütz. 2005. *Resource Use in European Countries: An Estimate of Materials and Waste Streams in the Community Including Imports and Exports Using the Instrument of Material Flow Analysis*. Project WP3c Cooperation with DG Environment—Thematic Strategy on the Sustainable Use of Resources. Available at <http://scp.eionet.europa.eu/themes/mfa/Zero%20Study> (accessed May 4, 2008).

Nandan, Satya. 2006. "The work of the International Seabed Authority and its relationship with the tribunal." Remarks at the International Tribunal for the Law of the Sea Tenth Anniversary, September 29. Available at <http://www.itlos.org/news/statement.nandan.final.version-acb01-02-07clean1.doc> (accessed May 4, 2008).

Nash, Gary. 2006. Opening speech at the Canada-China Mining Investment Forum, March 7. Hong Kong.

Norgaard, Richard B. 1990. "Economic indicators of resource scarcity: A critical essay." *Journal of Environmental Economics and Management* 19 (1): 19–25.

NMAB (National Materials Advisory Board). 2008. *Managing Materials for a Twenty-first Century Military*. Washington, DC: National Academy Press.

NRC (Committee on Critical Mineral Impacts of the U.S. Economy, Committee on Earth Resources, National Research Council). 2008. *Minerals, Critical Minerals, and the U.S. Economy*. Washington, DC.: National Academies Press. Available at <http://www.nap.edu/catalog/12034.html> (accessed April 4, 2008).

NRCan (Natural Resources Canada). 1996. *The Minerals and Metals Policy of the Government of Canada: Partnerships for Sustainable Development*. Available at <http://www.nrcan.gc.ca/mms-smm/poli-poli/pdf/mmp-eng.pdf> (accessed April 4, 2008).

NRCan (Natural Resources Canada). 2006. Canada-China Mining Investment Forum. March 7. Hong Kong.

NREL (National Renewable Energy Laboratory). 2004. "PV FAQ's: Will we have enough materials for energy-significant PV production." *U.S. Department of Energy Report* DOE/GO-102004-1834. Available at <http://www.nrel.gov/docs/fy04osti/35098.pdf> (accessed April 4, 2008).

O'Lear, Shannon. 2005. "Resource concerns for territorial conflict." *GeoJournal* 64 (4): 297–306.

Olsson, Ola, and Heather C. Fors. 2004. "Congo: The prize of predation." *Journal of Peace Research* 41 (3): 321–336.

Otto, James, and John Cordes. 2002. *The Regulation of Mineral Enterprises: A Global Perspective on Economics, Law, and Policy*. Rocky Mountain Mineral Law Foundation. Available at <http://www.rmmlf.org/pubs/mlp.pdf> (accessed May 4, 2008).

PAC (Partnership Africa-Canada). 2005. *Diamonds and Human Security*. Available at <http://pacweb.org/e/index.php?option=content&task=view&id=38&Itemid=61> (accessed May 14, 2008).

Pegg, Scott. 2003. "Globalization and natural-resource conflicts." *Naval War College Review* 56 (4): 82–96.

Rajan, Raghuram. 2006. "The great game again?" *Finance and Development* 43 (4): 54–55.

Reynolds, Douglas B. 1999. "The mineral economy: How prices and costs can falsely signal decreasing scarcity." *Ecological Economics* 31 (1): 155–166.

Robinson, Mary. 2006. Statement by Mary Robinson on human rights issues in Ghana's mining sector. November 22. Available at <http://www.realizingrights.org/pdf/MR_Statement_11-22-06_Oxfam_Mining_Ghana.pdf> (accessed May 30, 2007).

Robinson, James A., Ragnar Torvik, and Thierry Verdier. 2006. "Political foundations of the resource curse." *Journal of Development Economics* 79 (2): 447–468.

Ross, Michael. 2002. *Natural Resources and Civil War: An Overview with Some Policy Options*. Draft report prepared for the Governance of Natural Resources Revenues conference, sponsored by the World Bank and the Agence Française de Dévelopement, Washington, DC, December 9–10.

Ross, Michael. 2003. "The natural resource curse: How wealth can make you poor." In *Natural Resources and Violent Conflict: Options and Actions*, ed. Ian Bannon and Paul Collier, 1–37. Washington, DC: World Bank.

Ross, Michael. 2004. "What do we know about natural resources and civil war?" *Journal of Peace Research* 41 (3): 337–356.

Sachs, Jeffrey D., and Andrew M. Warner. 2001. "The curse of natural resources." *European Economic Review* 45 (4): 827–838.

Schauer, Tomas. 2007. "Minerals and limits to growth." Presentation at the Raw Materials Initiative Workshop INFRA 25708, Ljubljana, December 10–11.

Schultz, Klaus J., and Joseph A. Briskey. 2003. "The global mineral resource assessment project." USGS Fact Sheet FS-053-03. Available at <http://www.usgs.gov> (accessed April 4, 2008).

Secretariat for Equator Principles. 2006. *Equator Principles*. Available at <http://www.equator-principles.com/index.html> (accessed April 4, 2008).

Sen, Gautam. 1984. *The Military Origins of Industrialization and International Trade Rivalry*. New York: St. Martin's Press.

Shields, Deborah J., and Slavko V. Šolar. 2006. "The nature and evolution of mineral supply choices." In *Proceedings of the 15th International Symposium on Mine Planning and Equipment Selection*, ed. Mircea Cardu, Raimondo Ciccu, Enrico Lovera, and Enrica Helotti, 902–907. September 20–22, 2006, Torino, Italy. Galliate, IT: FIORDO.

Shields, Deborah J., Slavko V. Šolar, and William H. Langer. 2006. "Sustainable development and industrial minerals." In *Industrial Minerals and Rocks: Commodities, Markets, and Users*, ed. Jessica Kogel, Nikhil Trivedi, James Barker, and Stanley Krukowski, 133–142. Littleton, CO: Society for Mining, Metallurgy, and Exploration.

Shields, Deborah J., Slavko V. Šolar, and Michael D. Miller. 2007. "Mineral policy in the era of sustainable development: Historical context and future content." In *Proceedings of the 3rd International Conference on Sustainable Development Indicators in the Minerals Industry (SDIMI 2007)*, ed. Zacharias Agioutantis, 25–32. Athens: Heliotopos.

Simpson, David R., Michael A. Toman, and Robert U. Ayers. 2005. *Scarcity and Growth Revisited*. Washington, DC: Resources for the Future.

Singer, Peter W. 2003. *Corporate Warriors*. Ithaca, NY: Cornell University Press.

Skinner, Brian J. 2001. "Exploring the resource base." Presentation at the Long Run Availability of Minerals workshop, Washington, DC, April 22–23. Available at <http://www.rff.org/rff/Documents/RFF-Event-April01-keynote.pdf> (accessed April 4, 2008).

Slade, Margaret. 1982. "Cycles in natural-resource commodity prices: An analysis of the frequency domain." *Journal of Environmental Economics and Management* 9 (2): 138–148.

Smith, Kerry V. 1980. "The evaluation of natural resource adequacy: Elusive quest or frontier of economic analysis?" *Land Economics* 56 (3): 257–298.

Svedberg, Peter, and John E. Tilton. 2006. "The real, real price of nonrenewable resources: Copper 1870–2000." *World Development* 34 (3): 501–519.

Switzer, Jason. 2001. "Armed conflict and natural resources: The case of the minerals sector." Discussion paper. Available at <http://www.iisd.org/pdf/2002/envsec_mining_conflict.pdf> (accessed April 4, 2008).

Tilton, John E. 2003. "On borrowed time? Assessing the threat of mineral depletion." *Journal of Economic History* 63 (2): 612–613.

Tilton, John E., and Graham Davis. 2008. "Why the resource curse is a concern." *Mining Engineering* 60 (4): 29–32.

Tilton, John E., and Gustavo Lagos. 2007. "Assessing the long-run availability of copper." *Resources Policy* 32 (1–2): 19–23.

UK (United Kingdom, Department for Environment, Transport, and Regions). 1999. *A Better Quality of Life: A Strategy for Sustainable Development for the United Kingdom.* London: Stationary Office.

UN (United Nations). 1992. Agenda 21: Report of the United Nations Conference on Environment and Development. Rio de Janeiro, June 3–14. New York: United Nations.

UN (United Nations). 2002. Report of the World Summit on Sustainable Development. UN Doc. A/Conf.199/20. New York: United Nations. <http://www.johannesburgsummit.org/html/documents/documents.html> (accessed April 4, 2008).

USDA FS (U.S. Department of Agriculture Forest Service). 2004. *National Report on Sustainable Forests: 2003.* FS-766. Washington, DC: Government Printing Office.

Varma, Somit. 2008. Keynote presentation at the World Mines Ministries Forum, Toronto, February 29. Available at <http://www.wmmf.org/downloads.html> (accessed April 4, 2008).

Wang, Dianzuo. 2005. "Perspectives on China's mining and mineral industry." In *A Review on Indicators of Sustainability for the Mineral Extraction Industries*, ed. Roberto C. Villas-Boas, Deborah Shields, Slavko V. Šolar, Paul Anciaux, and Güven Önal, 105–113. Rio de Janeiro: CETEM/MCT/ CNPq/CYTED/IMPC.

WBCSD (World Business Council for Sustainable Development). 2000. *Sustainable Development Reporting: Striking the Balance.* Geneva: World Business Council for Sustainable Development.

World Bank. 2003. *Extractive Industries Review: Volume I, Striking a Better Balance.* Available at <http://irispublic.worldbank.org/85257559006C22E9/All+Documents/85257559006C22E985256FF6006843AB/$File/volume1english.pdf> (accessed April 4, 2008).

V

Conclusion

11

Resource Scarcity and Environmental Degradation: Implications for the Development of International Cooperation

Shlomi Dinar

This volume has sought to investigate how resource scarcity and environmental degradation motivate interstate cooperation. As Paul Diehl and Nils Petter Gleditsch (2001, 4) have commented, "the environmental security field remains handicapped in its theoretical focus on only one side of the conflict-cooperation coin. . . . That is, almost all the claims focus on the environment's conflict generating capacity and ignore the cooperative elements that might be present." Similar assertions have been echoed by other scholars (Deudney 1999; Barnett 2000). In short, this volume contends that although environmental change may generally be a catalyst for interstate conflict (Tir and Diehl 1998; Hensel, Mitchell, and Sowers 2006; Gleditsch et al. 2006), scarcity and degradation may also be the impetus for cooperation and peace (Conca and Dabelko 2002). To that extent, such a focus is important for understanding and analyzing regional and international security and stability as well (Goldstone 2001).

In general, we examine the extent to which scarcity promotes interstate cooperation, across those resources thought to be most prone to violent interstate conflict and resource wars (oil, freshwater, minerals, and fisheries), but also those resources or environmental problems that are likely to instigate merely political conflict or tensions among states (climate change, ozone, biodiversity, transboundary air pollution, and oceans pollution). We further speculate that different levels of scarcity and degradation should matter when explaining instances of cooperation. In short, low and high levels of scarcity and degradation may dampen interstate coordination—the former, because there is little urgency for cooperation given the relative abundance of the resource, and the latter, because there is either little of the resource to divide among the protagonists or the degradation is so severe that it is highly costly to remedy. While scarcity and degradation are important explanatory variables in our investigation, each chapter considers additional variables that may motivate or inhibit cooperation in the face of scarcity. Finally, we conjecture that in analyzing and understanding international negotiation and cooperation, country asymmetries or differences (geographic, economic, and political) matter. Such asymmetries indirectly relate to a state's perception of scarcity or degradation. In turn, since asymmetries often exacerbate interstate tensions in transboundary environmental affairs, offsetting such lopsidedness is critical to encourage interstate coordination.

Major Findings and Implications

In all, we find support for the scarcity-cooperation contention yet compelling differences are discernible across several of the issue topics, particularly in relation to how different levels of scarcity and degradation may affect interstate coordination. Perhaps equally noteworthy are the cooperative-inducing characteristics of scarcity in the case of those resources that have been popularly associated with international armed conflict and war. Those respective chapters find not only evidence of cooperation, in those regions that some scholars maintain are vulnerable to violent conflict, but likewise introduce additional examples of interstate coordination to buttress their core argument. As expected, scarcity is not always sufficient for explaining trends in cooperation, as various topical issues boast their own set of auxiliary variables (such as domestic influences, political norms, structural constraints, and the sheer desire for stable markets). Country asymmetries also play a crucial explanatory role across all the transboundary issues investigated. While key for understanding the evolution of cooperation in their own right, such asymmetries frequently influence how the parties perceive scarcity and degradation. Varying scarcity levels and asymmetries also influence the type of cooperative regime formulated and the nature of cooperation, providing insights for both theory and practice.

For example, the topical issues investigated in the context of our global commons section revealed that under conditions of economic asymmetry among the protagonists, successful cooperation is often sought when the externality affects, or is believed to more seriously affect, the more developed country or coalition of countries. Since such countries have lower propensities to accept pollution (or scarcity), their ability to tolerate the externality is decreased. By extension, there is less urgency to deal with the externality through interstate coordination when the developed country or coalition of developed countries is less affected by the degradation, or perceives the resource to be relatively plentiful (either physically or in terms of its economic value). When interstate coordination does become a desired policy option in situations characterized by increasing scarcity, degradation, and country asymmetries, however, incentives generally have to be provided (usually from developed states that are more sensitive to the degradation to less developed states) to facilitate formal cooperation. Where interstate coordination indeed succeeded in these asymmetrical instances, side payments or other related inducements were frequently used to affect bargaining power. These particular findings largely hold across other transboundary environmental issues investigated in this volume. By implication, in instances where cooperation has either failed or has had a limited foothold, such incentives may be considered as a viable strategy to foment interstate coordination. The findings of each chapter are synthesized below with respect to the hypotheses of this volume.

Chapter 2, on climate change and global warming, finds that the stock of greenhouse gases in the atmosphere is currently insufficient to produce the type and degree of optimal universal cooperation envisioned by many analysts and policymakers. (Countries that are currently emitting a sizable amount of carbon actually benefit from doing little abatement.) In other words, only when the risks of climate change become more near term (as clean air becomes scarcer) will the noncooperative solution become increasingly unpalatable. There will in fact be additional incentives for an increasing number of countries to enter a cooperative solution, and the likelihood of cheating will decrease as the net gains from abatement mount. This finding may suggest that an immediate or near-term search for a global cooperative initiative to combat climate change may face continued challenges. The current global economic recession may also temper hopes for swift international action on climate change.

Focusing on the Kyoto Protocol, the chapter also considers the various country asymmetries complicating cooperation. First, since the protocol negotiators focused mostly on the cost of abatement, they concentrated much less on the damages from climate change impacts, which tend to be different across states and regions. The emitters (mid- to high-level countries) of greenhouse gases are wealthy and more powerful relative to the main victims (low-latitude countries). Since emitters of greenhouse gases are relatively less harmed by their own pollution, any future climate change agreement will depend on the emitting countries agreeing to their liability for what happens to the poor in the low-latitude countries. In line with the scarcity contention postulated in this volume, cooperation is further complicated due to the relatively low scarcity levels (or impacts) experienced in mid- to high-level countries in comparison to low latitude ones. Effectively, poor low-latitude countries may have to subsidize richer countries, encouraging them to engage in carbon abatement—a scenario that seems unlikely for equity and fairness issues. The alternative, of course, is for richer states to pay into a fund that supports the efforts of poorer states to deal with the impacts of climate change—discussed as part of the nonbinding Copenhagen Accord of 2009. While this would be a more equitable scenario, this effort will continue to be challenged by the sheer fact that richer states are (in the near term) experiencing reduced environmental impacts, or scarcity, due to climate change. Such a dynamic between polluters and victims will continue to create a special problem for any future climate treaty.

Differences in levels of economic growth and development, for example, are also relevant to this discussion. The emission reductions targets stipulated in the Kyoto Protocol would largely benefit slow-growing countries (such as those in Western Europe) and hurt countries that were rapidly growing (such as the United States and China), since the latter group would need to abate a substantially higher percentage of pollution and bear a larger portion of the costs. Undoubtedly this provided a disincentive for certain developing and fast-growing countries to join the treaty. Varying levels of development also meant that the issue of tradable permits in the Kyoto Protocol favored Organization for Economic Cooperation and Development countries over developing countries. The protocol was based on actual emissions and effectively favored distributing tradable permits in proportion to global GDP rather than per person. The latter would be more amenable to developing countries, which account for the great majority of the global population.

Despite these obstacles, the chapter finds that the large increase in temperatures expected in the second half of this century will likely be more harmful (relative to the first half of the century) and may consequently encourage more comprehensive interstate coordination. The chapter also concludes that any future treaty must equate the marginal cost of abatement across sources across all countries in each time period (through efficient emission targets over time, permit trading, and carbon sequestration).

Chapter 3, on ozone, finds that the cooperative regime (the Montreal Protocol of 1987) resulted despite scientific uncertainty with regard to the extent and causes of ozone depletion. In line with our hypothesis, though, interstate cooperation broadened and deepened as evidence of ozone degradation increased (and as the protections given by the ozone layer were knowingly becoming scarcer). The London and Copenhagen conventions—of 1990 and 1992, respectively—were effectively amendments and adjustments to the original Montreal Protocol, given the new scientific findings and economics concerning ozone depletion.

The influence of industry and other domestic variables (such as public opinion and environmental groups), initially in the United States and subsequently in Europe, was also paramount in the evolution of cooperation. Equally important, however, were the inherent asymmetries in the ozone case and the manner in which they were offset. The difference in urgency witnessed in the ozone versus the climate change case may be partly explained by these inherent asymmetries.

In particular, as the science of ozone became more certain, the impacts became increasingly clear as well. Just after the Montreal Protocol was signed, research showed that the worst effects from ozone depletion can be found near Antarctica and the Arctic, effectively bearing on developed countries in the Northern Hemisphere as well as a mix of developed (e.g., Australia) and middle-income developing (e.g., Chile and Argentina) states in the Southern Hemisphere. Several years after this finding there was additional evidence of an increase in harmful UV-B radiation in Europe and Australia, detection of ozone loss in the summer months in regions such as North America, and a higher estimate of U.S. deaths as a result of skin cancer. In all, and given their greater vulnerability (and the would-be payoffs they would derive from abatement), it is not surprising that developed countries were relatively more concerned about ozone depletion, which may also explain the leadership (particularly as portrayed by the United States) that emerged just prior to the Montreal Protocol and, more important, the subsequent deepened cooperation that dealt with the ozone depletion problem. Combined with their shorter shadow of the future, developing countries, which were essentially less affected by ozone depletion, were able to employ particular bargaining strategies. Such countries claimed that until a mechanism for transferring funds from developed to developing countries was set up, they would not sign the protocol. Soon after this, the Montreal Protocol Multilateral Fund was created, and countries such as China and India became signatories. Incentives such as technology transfer and side payments were critical to offsetting the asymmetries in the ozone case, and were more in line with fairness and equity principles. Other incentives were likewise used to encourage full participation. One in particular constituted the primary enforcement mechanism. By banning trade between signatories and nonsignatories in chlorofluorocarbon substances, the treaty effectively controlled trade leakage and deterred free riding. In all, it was the strategy of positive inducements (including side payments and technology transfers) together with the enforcement strategy of trade restrictions that produced both a stable and fair agreement in this asymmetrical context.

Interestingly, the Montreal Protocol continues to be an inspiration for a climate regime. Yet the effects of the above-enumerated asymmetries (and their relationship to a country's perception of scarcity and degradation levels) on the evolution of cooperation in both cases indicate that these two global commons require different management institutions. Given the Kyoto Protocol's failures and the disappointing Copenhagen talks in 2009, policymakers and practitioners may wish to heed these differences as they go about designing a more successful governing institution. As this volume goes to press, the next major round of talks (COP 16) is set to commence in November 2010.

Chapter 4, on global biodiversity, finds that while early, indirect, and piecemeal efforts targeted at biodiversity protection are evidenced since the early 1900s, the Convention on Biodiversity (CBD), concluded in 1992, constituted the chief response to systematic and increasing biodiversity loss. In addition to the motivation supplied by sheer biodiversity loss, the chapter finds that international cooperation through the CBD was also motivated by the ecosystem services provided by biodiversity and the value of genetic resources to biotechnology and bioprospecting. In other words, cooperation was likewise facilitated because the increasing degradation and scarcity of biodiversity would compromise those ecosystem services along with the economic value of genetic resources.

The chapter also finds that evolving norms, a function of proliferating environmental attention and concern at the time, also facilitated the CBD. Still, the inherent country asymmetries, which subsequently had to be offset, also played a salient role in the evolving cooperation. Specifically, the majority of species diversity predominates in developing countries while the main entities with the ability to exploit such resources are the companies and countries of the developed world. Many of the powerful developed countries thus sought to obtain an international conservation agreement to protect the biological resources to which they wanted to maintain access. Conservation costs, however, would place a high burden on developing countries, and would have harmful effects on their agriculture and forestry industries. Moreover, plans by the developed world would essentially deny developing countries revenues to an equitable share of the genetic resources that they effectively owned. Since any successful agreement would necessitate the participation of developing countries, their bargaining power was subsequently affected. Perhaps most noteworthy are the financial resources and technology transfers to which the developed world committed to in the agreement.

Yet the chapter finds that the implementation of the CBD has faced particular stumbling blocks, especially as it pertains to its access and benefit sharing component, which was an important element of the developing countries' bargaining position as well. It is now unclear whether the gains to be made from bioprospecting deals (i.e., the value of genetic resources) will ever add up to the expectations, let alone meet the need for funding conservation. The argument is somewhat consistent with various economic studies that claim that the economic value of biodiversity is actually rather low because it remains relatively plentiful compared to the demand for it. This value will continue to decline so long as there is an ability to generate substitutes through synthetic chemistry and more sensitive screening. In the vernacular of our scarcity-cooperation contention, biodiversity has effectively been made less scarce given the decline in its value, which consequently has affected the CBD's implementation. That being said, chapter 4 affirms that the value of biodiversity (or the need for natural genetic resources) is likely to increase given continued concern with infectious diseases and the environmental soundness of natural resources compared to synthetic chemicals. This, in turn, may provide the urgency needed to successfully implement the CBD's access and benefit sharing component.

The issue topics investigated in various regional contexts provided a similar pattern of results regarding our theoretical contentions. Chapter 5, on transboundary air pollution, considers the formal cooperative regimes to combat air pollution established in Europe and North America, yet absent in East Asia. In fact, East Asia has witnessed only limited cooperation in the form of joint scientific research and monitoring activities in addition to some bilateral assistance programs. The chapter therefore finds that the scarcity of clean air has been a catalyst for cooperation, although inconsistently across the cases. It alludes to the severity of air pollution across East Asia, originating in China, and the associated cleanup costs (that would presumably have to be assumed by the principle victims of such pollution—South Korea and Japan) as a possible deterrent to formal cooperation. In other words, the serious impacts of severe transboundary acid rain (or the high degradation and scarcity of clean air) across East Asia, compounded by the associated abatement bill, acts as an effective deterrent to formal transboundary cooperation, at least in the short term.

Yet the evolution of cooperation in North America and Europe further sheds light on our scarcity contention, and highlights other critical explanatory variables such as domestic influences and country asymmetries. In both regions, the upwind/downwind dynamic pitted polluters against victims. Initially, polluting upwind countries, which were less affected by their own pollution (e.g., the United States in North America as well as the United Kingdom and Germany in Europe), supported only limited joint scientific research efforts and challenged their neighbors' calls for an abatement regime. It was only after such countries began to suffer recognizable ecological damages (and deal with increased clean air scarcity) that a more robust regionwide pollution abatement regime was considered. In fact, with growing evidence of damages in polluting states, domestic actors (NGOs, proenvironment political parties, the media, and the public) played a particularly notable role in demanding political action.

While a similar type of upwind/downwind air pollution dynamic is characteristic of East Asia, domestic forces lobbying in favor of abatement are all but absent. In China, where the effects of air pollution are severe, a closed and restrictive political regime hinders any efforts by Chinese communities to join with Japan and South Korea in calls for pollution abatement. China's shorter shadow of the future, compared to that of Japan and South Korea, also moderates its relative concern for pollution abatement, hindering formal cooperation in the region.

Chapter 6, on oceans pollution, argues that scarcity in clean seawater motivated the institutional organization of the Mediterranean Action Plan (MAP) in the mid-1970s. In fact, oceans pollution was not an issue of conflict or dispute but rather an opportunity for cooperation from the outset. Yet when the first results of the scientific research base indicated that pollution in the Mediterranean Sea tended to remain fairly local and did not generate a transboundary problem (or bathtub effect), the urgency for common and coordinated action was reduced. In line with the scarcity-cooperation contention of this volume, the finding seems to be one of the main reasons why there has been no perceived need for joint and formal environmental action in the Mediterranean case. Scarcity or degradation of the ocean, in other words, as it is regarded in a mutual or transborder sense, was negligible and thus did not necessitate interstate coordination on an environmental policy level. Some of the larger political conflicts in the region, combined with ideological and economic asymmetries among the protagonists, have further inhibited cooperation through MAP.

That being said, the chapter finds that cooperation via MAP since its inception has largely taken place on the scientific level. Cooperation therefore still unfolded, but on the normative scale—due mostly to the lack of knowledge in dealing with such localized pollution. To put it another way, cooperation among the Mediterranean countries was based on the related environmental problems they were experiencing rather than on a mutual dependence on each other for cross-border solutions to such pollution. Cooperation to this day continues to focus on normative issues and knowledge generation. MAP is also characterized as an enabling institution, getting particular in-state projects off the ground. By extension, MAP has also promoted liberal economic development and ecological modernization, hoping that such technological innovation and economic cooperation will bring about environmental improvement.

Turning to the issue topics that have been most popularly discussed in the context of the resource wars theory, we find further evidence for our scarcity-cooperation contention.

Chapter 7, on fisheries, finds that the common policy responses to international fisheries degradation include inaction, escalation, and negotiation, which can yield three categories of outcomes including further degradation, conflict, and cooperation. Interestingly, the chapter reveals that cooperation that is not preceded by escalation and conflict can often be less useful in addressing degradation than cooperation that results from conflict.

In addition, the chapter finds that the varying shadows of the future (and bargaining power) of the parties vis-à-vis the management of the fisheries resource play an important role in determining whether escalation or negotiation—and in turn, conflict or cooperation—will take place. A country's shadow of the future and bargaining power is determined by the relative wealth of the protagonists, the dependence of the national fisheries industry on a particular stock along with the size and range of the fishing fleet (substitutability), the directionality of the resource, and the political makeup of the respective entities. To that extent, perceived scarcity and degradation levels are largely a function of these inherent differences and asymmetries.

The chapter argues that cases that fall under the policy response of negotiation pertain to countries that generally share similar shadows of the future vis-à-vis the resource's management. In this instance, the chapter finds that cooperation is usually the first response to the onset of degradation. Agreement by members of the Convention for the Conservation of Antarctic Marine Living Resources (CCAMLR) on quotas for the fish stock shortly after the large-scale exploitation of Patagonian toothfish began is one example. In this context, parties perceived scarcity in a similar fashion.

The chapter assigns the term *escalation* to cases that include countries with varying shadows of the future. In this context, cooperation may not be the initial response to the onset of degradation since scarcity is lopsided. In effect, the state or group of states with the longer shadow of the future often has to employ either negative or positive incentives to encourage the other parties to cooperate. The so-called Turbot War, for example, between Canada and Spain highlights the use of such strategies. The disagreement between the two countries was a function of a higher unilateral fishing quota that was declared by the European Union (EU) on behalf of Spain. Although both parties were wealthy countries, Spain's shadow of the future was much shorter than that of Canada. In particular, while Canada had a near-shore fishery (designed to fish within Canada's exclusive economic zone and able to fish only a little farther out than its border), Spain's fishing fleet was a deep-sea one. As such, Spain was not concerned with the unsustainable management of turbot, relative to Canada, which was a result of the higher quota. In other words, the degradation was perceived to be quite low by Spain and hence did not warrant immediate action. Concerned by the degradation, Canada gave itself the right to impound vessels fishing in international waters outside its Atlantic exclusive economic zone and subsequently impounded a Spanish trawler. While utilizing negative incentives, Canada engaged in diplomatic measures or positive incentives, which ultimately resulted in an improved management agreement between the two sides.

Escalation was also employed in the Patagonian toothfish example introduced above. In this instance, members of CCLAMR, which shared a long shadow of the future, were faced with fishers flying the flags of nonmembers that did not follow quota rules or were not part of any regional fisheries management organization with set fishing standards. These fishers were, in short, flying the flags of countries that had short shadows of the future. To encourage participation in the management of the fisheries in this context of asymmetry, CCLAMR members used market forces to discourage trade leakage by prohibiting the landing or importation of undocumented fish. Since it was unprofitable to fish in this manner, the incentive was created for fishers to either flag with a member of CCAMLR or fish elsewhere.

Chapter 8, concentrating on freshwater, joins other academic exercises and challenges popular prophecies of future water wars. In fact, if history is at all a lesson for the future, it is striking that the past has witnessed no such violent confrontations. The only recorded water war took place in 2500 BCE in modern-day Iraq. Even military skirmishes over water have been few and far between, and perhaps unsurprisingly, have been limited to the arid Middle East. Political conflicts, however, have been common in the area of hydropolitics, but treaties to resolve such disputes have been equally common. This is the case not only for notoriously difficult issue topics (such as water quantity and allocation) but also for those issues generally regarded as relatively less contentious (such as hydropower, flood control, and pollution abatement). The chapter highlights the Indus River Agreement of 1960 as an example of the former scenario and the Rhine River Agreement of 1976 as an instance of the latter situation.

Empirical studies have also considered the role of scarcity in fostering formalized cooperation in the form of an international water treaty. The chapter looks at two findings of various empirical studies. While one finds support for our general scarcity-cooperation contention, the other underscores our secondary scarcity assertion pertaining to levels of degradation. In other words, while increasing scarcity motivates cooperation, high and low water scarcity levels moderate the likelihood of treaty formation. This second finding may be particularly significant, since voluntary cooperation is expected to be least likely to emerge in these instances. In this case, outside agents such as regional powers or international institutions may be instrumental in providing financial incentives or investment opportunities, or in simply promoting benefit-sharing initiatives as a means to encourage interstate coordination

While the chapter also considers how domestic variables may either inhibit (e.g., the case of India-Bangladesh hydropolitics) or promote (e.g., the case of Mexico–United States hydropolitics) cooperation, it highlights the role of country asymmetries as important for understanding the evolution of international cooperation. The upstream/downstream nature of international rivers combined with the relative power dynamics may make cooperation more challenging. Cooperation is perhaps most difficult when the upstream state is likewise the basin's hegemon (as in the case of the Tigris-Euphrates Basin, for instance) and is facilitated when the hegemon is downstream (as in the case of the Columbia River Basin, for example). In both cases, though, asymmetries were offset using strategies such as issue linkage and side payments. While the scarcity levels, compounded by the riparian asymmetries, were lopsided cooperation, still took place.

In particular, after successfully linking the Kurdish issue with water allocations on the Euphrates River, Syria and Turkey signed a protocol in 1987, guaranteeing Syria a set amount of water from Turkey. In the Columbia River Basin case, the United States not only had to compensate Canada as an incentive to join the hydropower and flood-control-based agreement but also had to cover a portion of the construction costs for upstream storage reservoirs. Canada also benefited from half the downstream power produced in the United States, as stipulated in the agreement of 1961 and its subsequent amendment in 1964. Additional examples of cooperation, which illustrate these two geographic and relative power associations, across pollution, flood control, hydropower, and water-quantity issues, are provided in the chapter.

Chapter 9, on oil, finds that the regions argued to be most prone to violent conflict (i.e., those that possess an impressive supply of recoverable fossil fuels and territorial contestation) such as the Persian Gulf, Caspian Sea, and Pacific Rim are likely not predestined to succumb to pessimist predictions at all. This is not to say that political tensions and disagreements are not witnessed, or that major regional powers refrain from engaging in some type of strategic maneuvering. Yet with a period of increasing oil scarcity, the chapter finds both instances of cooperation and peaceful efforts by the protagonists to define their share of the resources. In particular, the three regions represent distinct stages of oil development and stability. While the oil politics of the Gulf are the most stable and developed among the three regions—and demonstrate relative interconsumer and consumer-producer cooperation throughout history—those of the Caspian and Pacific Rim are seemingly the least mature, whereby lines of demarcation regarding the oil resources are yet to be drawn and where ownership issues remain to be settled. Even in these latter two cases, however, ecocynic prognoses are challenged.

Focusing on the superpower interests and rivalry in the Gulf throughout history (1945, 1973, 1979, 1991, and 2003), the chapter considers archival evidence to suggest that even in these select tense times, violent confrontation over the region's oil resources was never regarded as a serious option. The oil embargo in 1973, for example, rather solidified the notion among importing nations that the true divisions were between consumer and producer states rather than between consumer states. Likewise in 1991, the efforts of Saddam Hussein to create divisions among the consumer powers did not work as intended. Great powers like Russia and China did not come to Saddam's aid or even block the U.S.-led coalition. As in 1973, a consumer-producer dynamic was highlighted. Interconsumer relations in the Gulf have stayed relatively peaceful.

In the case of the Caspian Sea, the chapter finds that the main contentions among the regional producer nations—namely, the location of the oil pipelines and the legal status of the sea—have either led to agreement, in the case of the former issue (e.g., the BTC pipeline and several other agreements on export routes), or prompted a set of discussions, thus bridging the gap between the positions of various riparians. The chapter further shows that in both instances, the parties are well aware that conflict will only impede the maximum exploitation of the region's resource. The interests of outside powers also seem to overlap rather than conflict in the Caspian context. All major powers seem to be in agreement that if Caspian oil is not brought to market, than none of the countries will benefit. Such stable supply can only be secured if the superpowers refrain from meddling in the region's petropolitics or fomenting instability.

Similarly, the chapter finds that in the Pacific Rim, the three regional players (China, South Korea, and Japan) know very well that it is only the peaceful resolution of ownership issues of the South China Sea (and the associated oil resources) that will attract potential investors and energy companies. Instability will only drive such actors away. It is for this reason perhaps that we have witnessed the signing of a "code of conduct" between China and its fellow Association of Southeast Asian Nations with an interest in the various South China Sea disputes. Already in the East China Sea, China and Japan have worked to develop a conflict-avoidance regime to address their many claims. Unique to China, the chapter concludes that Beijing's preference for economic considerations (combined with a new leadership) over nationalistic imperatives challenges pessimistic accounts speculating that China's rise in the region will be of a violent nature.

Chapter 10, on minerals, considers both actual scarcity and impending perceived scarcity as important for understanding conflict and cooperation. Using this framework, it identifies several alternative dimensions of scarcity, beyond the known physical scarcity. These alternative forms of scarcity include: situational and locational scarcity (a consequence of a lack of investment in the discovery of new deposits to replace those being depleted), political scarcity (a result of interstate strategic rivalry and intrastate violence in an exporting country), and social scarcity (due to the perceived high costs of mineral extraction and processing by an exporting country, and the environmental impacts, community dislocation, and unfair treatment of workers given mineral extraction).

In explaining the conflicts that have arisen over minerals in modern times, the chapter claims that alternative forms of mineral depletion such as locational, political, and social scarcity have played a pivotal role. In fact, no interstate conflicts in recent years have been driven by sheer mineral depletion—although this situation could change if the demand for commodities rises dramatically again. In response to these different scarcity scenarios, efforts to define cooperative solutions have gained momentum in recent years. Market-, policy-, and consensus-based incentives highlighted by the chapter coincide with our scarcity-cooperation contention. In particular, they relate to the various scarcity dimensions enumerated above.

Like the chapter on oil, chapter 10 confirms that market-based incentives provide a great impetus for internationalized cooperation among states and firms in the minerals sector. For one, high mineral prices have motivated governments and firms to acquire commodity producers; the justification for such actions is economic security. High levels of demand in addition to the desire to create a profitable minerals sector have had the effect of bringing together various stakeholders (such as states and firms) for development projects. Agreements between consuming and producing countries clearly fall under this rubric. To that extent, China's cooperative stance vis-à-vis Africa, Australian-Japanese cooperation, and Chinese-Canadian cooperation are notable. The Global Mineral Resource Assessment Project constitutes an effort between producer and consumer countries to assess the exploration as well as development of the world's undiscovered nonfuel mineral resources. Agreements among producing nations are also relevant to this specific case, with the Association of Southeast Asian Nations' Minerals Cooperation Action Plan of 2005–2010 in East Asia and the Mines Ministries of the Americas in the Western Hemisphere constituting two major examples of regional cooperation. Other avenues for cooperation in light of market-based incentives are apparent through various intergovernmental organizations including the Integrated Commodity Program, an outgrowth of the United Nations Conference on Trade and Development, the Integrated Programme for Commodities, and the International Seabed Authority, an outgrowth of the United Nations Convention on the Law of the Sea.

Policy-based incentives also play a critical part in promoting cooperation in the mining case. Sustainable development in mineral resources, for instance, has become a pillar of international cooperation, with countries like Canada, Germany, the United Kingdom, and India instituting such policies into their international mineral agendas. By extension, the issue of corporate social responsibility (CSR) is also relevant. State-owned and private firms, and thus consuming countries, promote CSR for the sole pragmatic reason that it is good business as it increases long-term growth and profits, and decreases the potential for internationalized conflict with the host government and community. Consequently, stakeholders must cooperate with host governments and communities to implement CSR. The European Union, in particular, largely made up of mineral consuming nations, has been hard at work to develop a unified raw materials policy requiring cooperation not only among concerned European nations but also among the relevant mineral-exporting countries. Ensuring sustainable and more transparent supply of minerals from third countries has become an important element of this European policy. International cooperation in this instance is also evidenced through intergovernmental bodies such as the Intergovernmental Forum on Mining, Minerals, Metals, and Sustainable Development, an outgrowth of the World Summit on Sustainable Development of 2002.

Finally, consensus-based incentives relate especially to what the chapter coins political scarcity and the associated intrastate violent conflicts over mineral rents. These in-country affairs are internationalized to the extent that foreign governments are invariably involved. These events are further internationalized to the degree that such internal conflicts could well disrupt production and lead to the price increases of the respective mineral, subsequently affecting consuming nations. Both from an economic and altruistic perspective, such conflicts can only be prevented or halted through international cooperation. This has resulted in a number of multistakeholder collaborative initiatives including the Extractive Industries Transparency Initiative and the Kimberley Process.

Final Thoughts

In light of the environmental security literature that often touts scarcity and degradation as catalysts for interstate conflict, we find support for our scarcity-cooperation contention across the issue topics investigated in this volume. In turn, such cooperation may serve to enhance regional and international security broadly defined (Soroos 1994, 329). For example, in the context of oil and minerals, which have been singled out among nonrenewable resources as likely to ignite interstate war, our findings paint a more optimistic scenario. Scarcity and the desire to harness such lucrative assets encourage countries to coordinate their respective actions. This does not mean that conflicts of interest are absent. Rather, it may mean that the prospects of violent interstate encounters over such scarce resources are diminished.

Yet some of the transboundary environmental cases we investigate also highlight how different levels of scarcity and degradation may inhibit the emergence of cooperation, and consequently prolong the mismanagement of the resource or dispute among the protagonists. This does not mean that cooperation cannot be fostered in these situations (as the case of ozone clearly demonstrates). Voluntary cooperation may instead be further challenged given the reduced urgency for action, by some or all of the parties, due to actual or perceived low levels of scarcity, or the sheer dwindling value of the resource. This was demonstrated most prominently in the case of climate change and freshwater, and to some degree, in the context of biodiversity, oceans pollution, transboundary air pollution, and fisheries. High scarcity and degradation levels were also found to inhibit cooperation in some of the cases due to the mere paucity of the resource or the high abatement costs associated with resolving the environmental externality. This was underscored most significantly in the instance of freshwater and, to a certain extent, transboundary air pollution.

While other variables are important for explaining the emergence of cooperation, country asymmetries are noteworthy for our particular analysis. Asymmetries not only reflect a given country's perception of scarcity and the urgency to cooperate but can likewise play an independent inhibiting role in state-to-state interactions in transboundary environmental affairs. Asymmetries also affect a country's bargaining power in negotiation. Since transboundary environmental problems often emerge between or among asymmetrical parties, understanding how cooperation may be fostered by offsetting such differences through, say, treaty design may be potentially instructive. Such a treaty could prove to be not only more fair and equitable but more stable and self-enforcing as well (Barrett 2003). This analysis therefore demonstrates that while levels of scarcity are in fact crucial for understanding the development of cooperation, so is the design of the institution used to foster formalized cooperation. Efficient treaties, in turn, may help promote regional and international stability and security.

References

Barnett, Jon. 2000. "Destabilizing the environment-conflict thesis." *Review of International Studies* 26 (2): 271–288.

Barrett, Scott. 2003. *Environment and Statecraft: The Strategy of Environmental Treaty Making*. Oxford: Oxford University Press.

Conca, Ken, and Geoffrey Dabelko, eds. 2002. *Environmental Peacemaking*. Washington, DC and Baltimore: Woodrow Wilson Center Press and Johns Hopkins University Press.

Deudney, Daniel. 1999. "Environmental security: A critique." In *Contested Grounds: Security and Conflict in the New Environmental Politics*, ed. Daniel Deudney and Richard Matthew, 187–223. Albany: State University of New York Press.

Diehl, Paul, and Nils Petter Gleditsch. 2001. "Controversies and questions." In *Environmental Conflict*, ed. Paul Diehl and Nils Petter Gleditsch, 1–9. Boulder, CO: Westview Press.

Gleditsch, Nils Petter, Kathryn Furlong, Håvard Hegre, Bethany Lacina, and Taylor Owen. 2006. "Conflict over shared rivers: Resource scarcity or fuzzy boundaries." *Political Geography* 25 (4): 361–382.

Goldstone, Jack. 2001. "Demography, environment, and security." In *Environmental Conflict*, ed. Paul Diehl and Nils Petter Gleditsch, 84–108. Boulder, CO: Westview Press.

Hensel, Paul, Sara McLaughlin Mitchell, and Thomas Sowers. 2006. "Conflict management of riparian disputes." *Political Geography* 25 (4): 383–411.

Soroos, Marvin. 1994. "Environmental security and the prisoner's dilemma." *Journal of Peace Research* 31 (3): 317–332.

Tir, Jaroslav, and Paul Diehl. 1998. "Demographic pressure and interstate conflict: Linking population growth and density to militarized disputes and wars, 1930–1989." *Journal of Peace Research* 35 (3): 319–339.

Index

Global Environmental Accord: Strategies for Sustainability and Institutional Innovation
Nazli Choucri, series editor

Nazli Choucri, ed., *Global Accord: Environmental Challenges and International Responses*

Peter M. Haas, Robert O. Keohane, and Marc A. Levy, eds., *Institutions for the Earth: Sources of Effective International Environmental Protection*

Ronald B. Mitchell, *Intentional Oil Pollution at Sea: Environmental Policy and Treaty Compliance*

Robert O. Keohane and Marc A. Levy, eds., *Institutions for Environmental Aid: Pitfalls and Promise*

Oran R. Young, ed., *Global Governance: Drawing Insights from the Environmental Experience*

Jonathan A. Fox and L. David Brown, eds., *The Struggle for Accountability: The World Bank, NGOS, and Grassroots Movements*

David G. Victor, Kal Raustiala, and Eugene B. Skolnikoff, eds., *The Implementation and Effectiveness of International Environmental Commitments: Theory and Practice*

Mostafa K. Tolba with Iwona Rummel-Bulska, *Global Environmental Diplomacy: Negotiating Environmental Agreements for the World, 1973–1992*

Karen T. Litfin, ed., *The Greening of Sovereignty in World Politics*

Edith Brown Weiss and Harold K. Jacobson, eds., *Engaging Countries: Strengthening Compliance with International Environmental Accords*

Oran R. Young, ed., *The Effectiveness of International Environmental Regimes: Causal Connections and Behavioral Mechanisms*

Ronie Garcia-Johnson, *Exporting Environmentalism: U.S. Multinational Chemical Corporations in Brazil and Mexico*

Lasse Ringius, *Radioactive Waste Disposal at Sea: Public Ideas, Transnational Policy Entrepreneurs, and Environmental Regimes*

Robert G. Darst, *Smokestack Diplomacy: Cooperation and Conflict in East-West Environmental Politics*

Urs Luterbacher and Detlef F. Sprinz, eds., *International Relations and Global Climate Change*

Edward L. Miles, Arild Underdal, Steinar Andresen, Jørgen Wettestad, Jon Birger Skjærseth, and Elaine M. Carlin, *Environmental Regime Effectiveness: Confronting Theory with Evidence*

Erika Weinthal, *State Making and Environmental Cooperation: Linking Domestic and International Politics in Central Asia*

Corey L. Lofdahl, *Environmental Impacts of Globalization and Trade: A Systems Study*

Oran R. Young, *The Institutional Dimensions of Environmental Change: Fit, Interplay, and Scale*

Tamar L. Gutner, *Banking on the Environment: Multilateral Development Banks and Their Environmental Performance in Central and Eastern Europe*

Liliana B. Andonova, *Transnational Politics of the Environment: The European Union and Environmental Policy in Central and Eastern Europe*

David L. Levy and Peter J. Newell, eds., *The Business of Global Environmental Governance*

Dennis Pirages and Ken Cousins, eds., *From Resource Scarcity to Ecological Security: Exploring New Limits to Growth*

Ken Conca, *Governing Water: Contentious Transnational Politics and Global Institution Building*

Sebastian Oberthür and Thomas Gehring, eds., *Institutional Interaction in Global Environmental Governance: Synergy and Conflict among International and EU Policies*

Ronald B. Mitchell, William C. Clark, David W. Cash, and Nancy M. Dickson, eds., *Global Environmental Assessments: Information and Influence*

Helmut Breitmeier, Oran R. Young, and Michael Zürn, *Analyzing International Environmental Regimes: From Case Study to Database*

J. Timmons Roberts and Bradley C. Parks, *A Climate of Injustice: Global Inequality, North-South Politics, and Climate Policy*

Saleem H. Ali, ed., *Peace Parks: Conservation and Conflict Resolution*

D. G. Webster, *Adaptive Governance: The Dynamics of Atlantic Fisheries Management*

Shlomi Dinar, ed., *Beyond Resource Wars: Scarcity, Environmental Degradation, and International Cooperation*